ENVIRONMENTAL TOXICOLOGY

ENVIRONMENTAL TOXICOLOGY
Concepts and Principles

Mohd Salim Reshi | Darakhshan Javaid
Syed Sanober Qadri | Neelu Sinha

JENNY STANFORD
PUBLISHING

Published by

Jenny Stanford Publishing Pte. Ltd.
101 Thomson Road
#06-01, United Square
Singapore 307591

Email: editorial@jennystanford.com
Web: www.jennystanford.com

British Library Cataloguing-in-Publication Data
A catalogue record for this book is available from the British Library.

Environmental Toxicology: Concepts and Principles

ISBN 978-981-5129-68-7 (Hardcover)
ISBN 978-1-003-63707-3 (eBook)

This book is dedicated to all researchers, educators, and practitioners who tirelessly work to unravel the mysteries of toxic substances and their impact on life and the environment.

To my family, friends, and mentors, whose unwavering support and encouragement have been my foundation.

And to those countless individuals who have suffered the consequences of toxic exposures—may our work contribute to a safer, healthier future for all.

Contents

Prof. Sangeeta Shukla

CHAUDHARY CHARAN SINGH UNIVERSITY, MEERUT-250 004 (U.P.)
(NAAC A++ Accredited)

Professor Sangeeta Shukla
D.Sc.
Vice Chancellor

Foreword

Environmental toxicology is a critical field that merges the details of environmental science with the profound impact of toxic substances on living organisms. As our understanding of the environment evolves, so too does our recognition of the myriad ways in which human activities influence ecological and human health. This book, "Environmental Toxicology: Concepts and Principles," serves as a comprehensive guide, bridging the gap between fundamental toxicological principles and their practical applications in environmental contexts.

Over the past decades, the scope of environmental toxicology has broadened significantly. From the industrial pollutants that dominated early research to the more nuanced studies of chemical interactions and biological responses, our approach to understanding and mitigating the effects of toxic substances has become more sophisticated. This book provides an in-depth exploration of these advancements, offering readers a detailed understanding of the mechanisms by which toxicants interact with biological systems, the methodologies used to assess toxic effects, and the regulatory frameworks that guide environmental protection.

The authors have meticulously compiled a wealth of information, making complex concepts accessible to a wide audience. Whether you are a student embarking on your journey in environmental science, a researcher seeking to deepen your knowledge, or a policymaker aiming to craft informed regulations, this book will serve as an invaluable resource. The integration of case studies, contemporary research findings, and practical examples ensures that readers can appreciate the relevance of toxicological principles in real-world scenarios.

As we confront global challenges such as climate change, pollution, and biodiversity loss, the importance of environmental toxicology cannot be overstated. The insights gained from this field will be crucial in shaping sustainable solutions and protecting both human health and the environment. "Environmental Toxicology: Concepts and Principles" is not just a textbook; it is a call to action, encouraging readers to engage with the pressing environmental issues of our time and contribute to the creation of a healthier, more sustainable world.

Prof. Sangeeta Shukla

V.C. LODGE, UNIVERSITY CAMPUS, MEERUT - 250 004
Office : +91-0121-2760554, 2760551, Fax : 2762838
Camp Office : +91 - 0121-2600066, Fax : 2760577
web : www.ccsuniversity.ac.in
E-mail : vc@ccsuniversity.ac.in

Prof. Akbar Masood

Baba Ghulam Shah Badshah University

Prof. Akbar Masood
Ph.D; FASc(AW);FIBRF;FNESA;FISBT;FSTOX;FLS;ASTP;FHAS;FRSC; FABSc; FISROSET
Vice Chancellor

D.O. No: BGSBU/VC/24/1431

Dated : 06-08-2024

Foreword

In an era where environmental concerns are at the forefront of global discourse, the field of environmental toxicology has never been more pertinent. "Environmental Toxicology: Concepts and Principles" arrives at a crucial moment, offering a thorough examination of the scientific principles and methodologies that underpin our understanding of toxic substances in the environment.

This book is a testament to the interdisciplinary nature of environmental toxicology, seamlessly integrating knowledge from biology, chemistry, ecology, and public health. The authors have succeeded in crafting a narrative that is both comprehensive and accessible, ensuring that readers from diverse backgrounds can grasp the fundamental concepts and apply them to their respective fields. The chapters meticulously cover a wide range of topics, from the molecular mechanisms of toxicity to the assessment of ecological risks and the implications for human health.

As we navigate the complexities of environmental stewardship, the role of environmental toxicologists becomes increasingly vital. This book equips readers with the tools and understanding necessary to address environmental challenges with scientific rigor and ethical responsibility.

"Environmental Toxicology: Concepts and Principles" is more than an academic resource; it is an invitation to engage with the critical issues facing our planet. It encourages readers to think critically, act responsibly, and contribute to the advancement of environmental science. Whether you are a seasoned professional or a newcomer to the field, this book will inspire you to make meaningful contributions to the protection and preservation of our natural world.

Prof. Akbar Masood

Dhanore, Rajouri-Jammu & Kashmir (India) - 185234
Email :vc@bgsbu.ac.in II Website www.bgsbu.ac.in Telephone : +1962-241001 , Mobile : +9906966281

Preface

In crafting this book, we aimed to create a resource that is both informative and accessible, catering to a diverse audience that includes students, researchers, policymakers, and environmental professionals. We have endeavored to present the content in a clear and structured manner, ensuring that readers can easily navigate through the various topics and gain a holistic understanding of environmental toxicology.

The contents of this book are divided into several key sections, each focusing on different aspects of environmental toxicology. We begin with an introduction to the basic concepts, laying the groundwork for more detailed discussions on the mechanisms of toxicity, exposure assessment, and risk evaluation. Subsequent chapters delve into specific classes of toxicants, including heavy metals, pesticides, and industrial chemicals, examining their sources, pathways, and effects on ecosystems and human health.

One of the unique features of this book is the integration of contemporary research findings and case studies. These elements provide real-world context and demonstrate the application of theoretical knowledge to practical situations. By including these examples, we hope to illustrate the relevance of environmental toxicology in addressing current environmental challenges and to inspire readers to engage actively in this dynamic field.

We also recognize the importance of regulatory frameworks and policy considerations in the realm of environmental toxicology. Therefore, we have dedicated sections of this book to exploring the development and implementation of environmental regulations, the role of governmental and non-governmental organizations, and the ethical considerations that guide environmental protection efforts.

As the readers embark on their journey with this book, we hope that they find the content engaging, thought-provoking, and enlightening. Our goal is to equip them with the knowledge and tools necessary to understand and address the intricate challenges posed by environmental toxins. We invite them to explore the fascinating world of environmental toxicology and to join us in the collective effort to protect and preserve our environment for future generations.

Mohd Salim Reshi
Darakhshan Javaid
Syed Sanober Qadri
Neelu Sinha
January 2025

Acknowledgments

We are grateful to Almighty Allah for giving us the strength to complete this book successfully. It is indeed a great privilege for us to acknowledge our deep sense of gratitude to Prof. Sangeeta Shukla, Honourable Vice Chancellor, Chaudhary Charan Singh University, Meerut, UP, India, and Dr Sadhana Shrivastava, Senior Scientist Officer, Jiwaji University, Gwalior, MP, India, for their insightful guidance and firm support throughout this project. We extend our heartfelt thanks to Prof. Akbar Masood, Honourable Vice Chancellor, Baba Ghulam Shah Badshah University (BGSBU), Rajouri, J&K, India, as well as to the staff and students at BGSBU for offering the essential resources and a nurturing academic environment and valuable suggestions. We particularly acknowledge the Indian Council of Medical Research (ICMR) and the University Grants Commission (UGC) for their generous funding and support, which have been pivotal in enabling this project.

A special thanks to our families, teachers, colleagues, and friends for their unwavering support and encouragement throughout this project. Their patience and understanding have been essential in bringing this book to completion.

We are immensely grateful to our readers whose interest in environmental toxicology is a testament to the growing awareness and concern for environmental issues. We hope this book will serve as a useful guide and inspire many to make meaningful contributions to the field of environmental toxicology.

Thank you.

The Authors

Chapter 1

Introduction to Toxicology

"Decoding the science of poisons for a safer world"

1.1 Introduction

Even before the advent of civilization, man sought to devour a wide range of plant and animal-based things in his urge for sustenance. He eventually discovered that some of those compounds had negative consequences ranging from disease to death, while others had positive consequences. Foods were identified as chemicals that were useful and necessary for the survival and normal functioning of the body; whereas, poisons were identified as harmful substances. As a result, poisons have been known to man since the dawn of time.

As far as toxicology is concerned, toxicology is derived from the words: toxicum (toxicos or toxicon or toxin) or the late Latin word "**toxicus**," which means "**poison**," and "**logous**" (or logia), which means **study** or **knowledge** or **discourse**. The word **toxicus** is itself derived from the ancient Greek word **toxikon**, which means arrow dipped in deadly poison to make it lethal. Toxicology can broadly be defined as a fundamental science of poisons. Toxicology is the scientific discipline that studies the harmful effects of toxic substances on living organisms, as well as the mechanisms by

Environmental Toxicology: Concepts and Principles
Mohd Salim Reshi, Darakhshan Javaid, Syed Sanober Qadri, and Neelu Sinha

ISBN 978-981-5129-68-7 (Hardcover), 978-1-003-63707-3 (eBook)
www.jennystanford.com

which these substances exert their adverse effects. It involves the examination of how toxic agents interact with biological systems, the identification of potential health hazards and risks associated with exposure, and the development of methods to assess and mitigate these risks. Toxicology plays a crucial role in protecting human health, the environment, and other organisms by providing insights into the safety of drugs, chemicals, environmental pollutants, and various other substances. By convention, toxicology also includes the study of harmful effects caused by physical phenomena, such as radiation of various kinds and noise. In practice, however, many complications exist beyond these simple definitions, both in bringing more precise meaning to what constitutes a poison and the measurement of toxic effects. Poison is a quantitative concept, almost any substance being harmful at some doses but, at the same time, being without harmful effects at some lower doses. Between these two limits, there is a range of possible effects, from subtle long-term chronic toxicity to immediate lethality. **Vinyl chloride** may be taken as an example. It is a potent hepatotoxicant at high doses, a carcinogen with a long latent period at lower doses, and apparently without effect at very low doses. Clinical drugs are even more poignant examples because, although therapeutic and highly beneficial at some doses, they are not without deleterious side effects and may be lethal at higher doses. **Aspirin** (acetylsalicylic acid), for example, is a relatively safe drug at recommended doses and is taken by millions of people worldwide. At the same time, chronic use can cause deleterious effects on the gastric mucosa, and it is fatal at a dose of about 0.2–0.5 g/kg. Approximately 15% of reported accidental deaths from poisoning in children result from ingestion of salicylates, particularly aspirin.

Toxicology can also be defined as a field of science that helps us to understand the harmful effects that chemicals, substances, or situations, can have on people, animals, and the environment. Some refer to toxicology as the **"Science of Safety"** because as a field it has evolved from a science focused on studying poisons and adverse effects of chemical exposures to a science devoted to studying safety.

Broader definitions of toxicology, such as "the study of the detection, occurrence, properties, effects, and regulation of toxic substances," although more descriptive, do not resolve the difficulties. Toxicity itself can rarely, if ever, be defined as a single molecular event but is, rather, a cascade of events starting with exposure, proceeding through distribution and metabolism, and ending with interaction with cellular macromolecules (usually DNA or protein) and the expression of a toxic endpoint. This sequence may be mitigated by excretion and repair.

1.2 Scope of Toxicology

Toxicology, in its broadest sense, is one of the oldest fields of science. Man must have been aware of several poisons from his earliest days, such as snake venom and dangerous plant toxins. It is obvious from the oldest recorded texts that the ancients had a good understanding of poisons.

The contributions and activities of toxicologists are diverse and widespread. Toxicologists are interested in the mechanisms of action and exposure to chemical substances as a cause of acute and chronic sickness in the biomedical field. Toxicologists contribute to physiology and pharmacology by studying physiological processes using hazardous substances. They are involved in recognizing, identifying, and quantifying dangers resulting from occupational exposure to chemicals, as well as the public health aspects of chemicals in air, water, other areas of the environment, foods, and pharmaceuticals. Toxicologists have always played a key role in the research and development of novel medications and pesticides. Toxicologists also help set rules and laws that safeguard human health and the environment from the harmful effects of chemicals. Environmental toxicologists (a relatively new subsection of the subject) have broadened toxicology to examine the impact of pollutants on flora and fauna (a relatively young subsection of the subject). The methods through which toxicants influence cell proliferation and differentiation, as well as how cells respond to toxicants at the gene level, are being investigated by molecular toxicologists. Toxicologists study the processes by which substances cause harmful effects in biological systems in all fields of toxicology. Antidotes and treatment regimens for poisonings and xenobiotic damage are developed by clinical toxicologists. Toxicologists work in academic, industrial, and government institutions to do some or all these things. They share procedures for gathering data on material toxicity, as well as the duty of using that data to create acceptable forecasts about the material's dangers to humans and the environment. The science of toxicology is defined by these diverse but complementary activities.

1.3 Brief History of Toxicology

The origin of toxicology may be traced back to early cave dwellers who identified poisonous plants and animals and employed their extracts for hunting or warfare. Written records show that hemlock, opium, arrow poisons, and certain metals were employed to poison enemies or to carry out state executions around 1500 BC. With time, poisons became widely used and with great sophistication. Socrates, Cleopatra, and Claudius are among the famous poisoning victims. Certain key notions in toxicology began to take shape during the Renaissance and Age of Enlightenment. The studies of Paracelsus (~1500 AD) and Orfila (~1800 AD) are well known. Paracelsus determined that specific chemicals were responsible for the toxicity of a plant or animal poison. He also documented that the body's response to those chemicals depended on the dose received. His studies revealed that small doses of a substance might be harmless or beneficial; whereas, larger doses could be toxic. This is now known as the dose-response relationship, a major concept of toxicology. Paracelsus was one of the founders of modern toxicology. His best-known quote: "All things are poison, and nothing is without poison; the dosage alone makes it so a thing is not a poison." Orfila, a Spanish physician, is often referred to as the founder of toxicology. It was Orfila who first prepared a systematic correlation between the chemical and biological properties of poisons of the time. He demonstrated the effects of poisons on specific organs by analyzing autopsy materials for poisons and their associated tissue damage. The 20th century is marked by an advanced level of understanding of toxicology. DNA (the molecule of life) and various biochemicals that maintain body functions were discovered. Our level of knowledge of toxic effects on organs and cells is now being revealed at the molecular level and it has been recognized that virtually all toxic effects are caused by changes in specific cellular molecules and biochemicals.

Furthermore, with the publication of her controversial book, *The Silent Spring*, in 1962, Rachel Carson became an important influence in initiating the modern era of Environmental Toxicology. Her book emphasized stopping the widespread, indiscriminate use of pesticides and other chemicals and advocated use patterns based on sound ecology.

It is obvious, however, that toxicology has entered a period of fast development since the 1960s, transitioning from a purely descriptive discipline to one in which the relevance of hazardous action mechanisms is widely recognized. With a greater emphasis on the use of molecular biology techniques since the 1970s, the rate of change has accelerated even more, and important discoveries have been achieved in a variety of fields, including chemical carcinogenesis and xenobiotic metabolism, among many others.

1.4 Toxicology and Its Relations with Other Sciences

The scientific study of poisons or poisonous substances and how they affect living things is known as toxicology. It is essential to comprehend the possible dangers and risks related to various chemical, physical, and biological agents. With the goal of safeguarding public health and the environment, toxicologists study how toxic compounds interact with biological systems, the mechanisms of toxicity, and the factors that affect toxicity.

The broad field of toxicology has close ties to numerous other sciences (Fig. 1.1). The following are some of the main fields that toxicology interacts with:

1. Pharmacology: The sciences of toxicology and pharmacology are interconnected. The study of drugs and how they interact with the body, including how they are absorbed, distributed, metabolized, and eliminated, is the primary focus of pharmacology. The study of harmful effects brought on by exposure to drugs or other chemicals is known as toxicology. When evaluating the safety and effectiveness of drugs and other substances, an understanding of pharmacokinetics and pharmacodynamics is important.

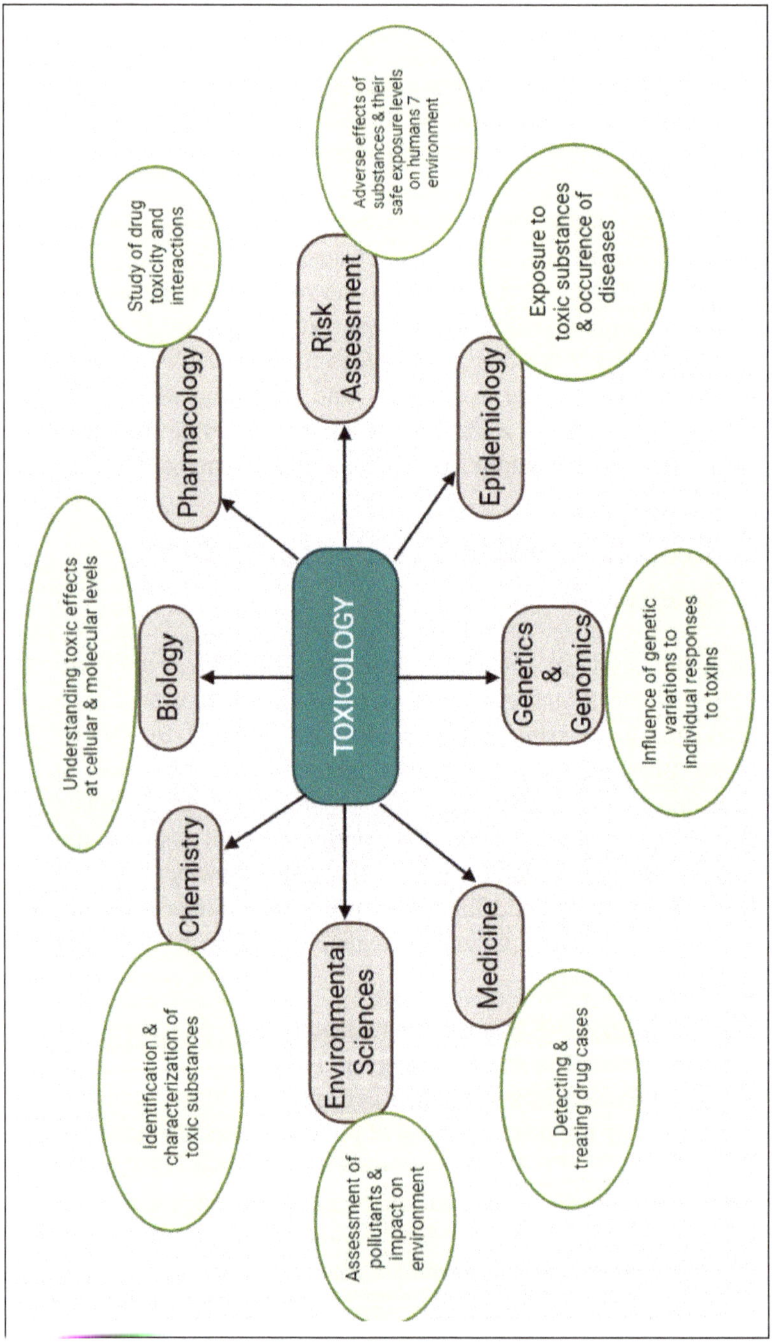

Figure 1.1 Showing toxicology as an interdisciplinary science.

2. Chemistry: Chemistry is essential to toxicology because it aids in the identification, specification, and quantification of toxic substances. Analytical chemistry techniques are used by toxicologists to detect and measure toxic chemicals in various environmental samples or biological tissues. To understand the behavior and potential effects of toxins, researchers also examine their chemical composition.

3. Biology: To fully understand how toxic compounds interact with living beings at the molecular, cellular, and physiological levels, toxicology strongly relies on biological sciences. This involves research on cellular responses, toxin metabolism, and toxic action mechanisms.

4. Environmental Science: Toxicology and environmental science are closely linked due to their shared concern for the impact of toxic substances on the environment. Environmental toxicology is concerned with how pollutants affect ecosystems, wildlife, and the environment as a whole. It facilitates the identification and mitigation of threats to ecological balance and biodiversity.

5. Medicine: Toxicology has significant applications in clinical medicine. Clinical toxicologists are experts in detecting and treating poisoning cases as well as adverse drug responses in humans. By providing advice on how to handle toxic exposures, they serve a crucial role in emergency care and public health.

6. Genetics and Genomics: The ability to examine how genetic variations affect individual responses to poisons has revolutionized toxicology because of advances in genetics and genomics. The scientific area of toxicogenomics identifies biomarkers of toxicity and vulnerability to certain toxins.

7. Epidemiology: Toxicology and epidemiology frequently work together to study the relationships between exposure to toxic substances and the occurrence of diseases or other negative health outcomes in human populations. Epidemiological studies can help establish causative links between environmental exposures and health effects.

8. Risk Assessment: Toxicology is an important component of risk assessment, which includes assessing potential adverse impacts of substances and figuring out safe exposure levels for both humans and the environment. To make educated decisions about regulatory policies and public health guidelines, risk assessment takes toxicological data along with exposure assessments into account.

Toxicology is an interdisciplinary science that integrates knowledge from numerous disciplines (Fig. 1.1) to comprehend the effects of toxic substances and develop strategies for controlling and reducing hazards to both human health and the environment.

Chapter 2

Toxicology: Disciplines and Importance

"Every facet of toxicology enhances our safety"

2.1 Disciplines of Toxicology

As modern toxicology evolved, a number of subdisciplines developed, each focusing on a particular concern (Fig. 2.1). The many subdisciplines collaborate to produce a powerful discipline committed to preserving public health through a greater knowledge of the adverse impacts of chemicals on biological systems much like the brass, strings, and percussion instruments of an orchestra.

2.1.1 Descriptive Toxicology

The foundation of toxicological knowledge, at its most basic level, is understanding the unique toxicities that accompany the exposure of a certain species to a given chemical. This type of information is collected by studying the model organisms after they have been exposed to highly regulated dosages of the chemical in question, with thorough records of the time, amount, and frequency of the compound administration. Such investigations are complicated and expensive, so it is no wonder that specialized firms and government institutes

Environmental Toxicology: Concepts and Principles
Mohd Salim Reshi, Darakhshan Javaid, Syed Sanober Qadri, and Neelu Sinha

ISBN 978-981-5129-68-7 (Hardcover), 978-1-003-63707-3 (eBook)
www.jennystanford.com

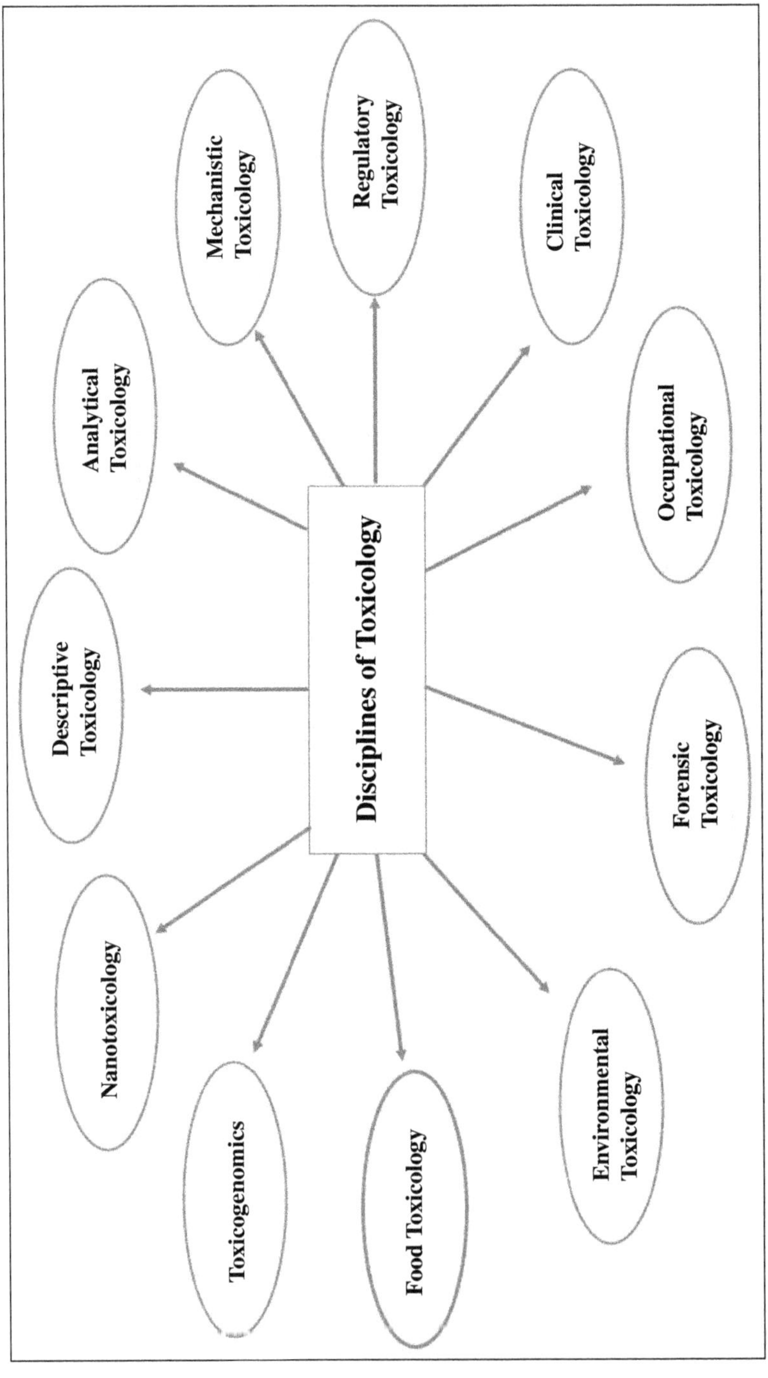

Figure 2.1 Different disciplines of toxicology.

have sprung up to do these types of toxicity evaluations. Because the well-being of the animals must be prioritized if high-quality data is to be acquired, certain care must be paid to inbred animal strains, bedding, cages, feed, and living circumstances, among other things. In these contexts, descriptive toxicologists are experts in the design of toxicity experiments employing a variety of model animals such as rats, mice, guinea pigs, dogs, primates, and even zebrafish or invertebrates. To evaluate novel compounds, a variety of animal-based models are utilized, ranging from experiments that look at short-term hazardous effects to those that emerge over months or even years of exposure (e.g., cancer). To assess the sensitivity of unborn animals to novel substances, tests on pregnant animals are also carried out. However, in vitro studies utilizing primary or grown cell lines are among the tools employed by descriptive toxicologists. Knowledge of comparable sciences including histology, anatomy, cell biology, and pathology is essential for descriptive toxicology. A solid foundation in the utilization of cell and tissue imaging technologies to identify the subtle effects of chemicals on the precise architecture of bodily systems is also beneficial in today's environment.

2.1.2 Analytical Toxicology

Even if a chemical is suspected of causing sickness or poor health in humans, rigorous confirmation is essential to ensure that toxicologically relevant amounts of the substance are present in the bodies or tissues of those who have been exposed. In humans, correlations between blood concentrations of a hazardous agent and the severity of pathological outcomes are invaluable if dose–response relationships for causality are to be established. Analytical toxicologists are experts in using sensitive detection equipment to test body samples acquired from exposed persons – most typically blood or urine samples – for the presence of suspected toxins. These techniques are now being utilized in industrial settings to identify the usage of illegal substances by workers. Such analytical abilities are especially critical when descriptive toxicologists evaluate medications and toxic chemicals in lab animals. Analytical data from these tests can indicate how much of a chemical is absorbed into the circulating bloodstream and tissues of exposed animals via various routes of exposure (e.g., when the chemical is administered via the diet, drinking water, or inhaled air of lab animals). Analytical toxicology as a subdiscipline leans heavily on the knowledge and trust offered by modern analytical chemistry; as a result, a solid understanding of organic and inorganic chemistry is necessary.

2.1.3 Mechanistic Toxicology

The presence of toxic reactions to a chemical in an animal test does not always mean that the substance poses a risk to human health when used normally. Sometimes, if high dosages of chemicals are given to rats or mice, adverse responses can arise that do not occur in humans who are exposed to lower doses. Indeed, toxicological history is littered with health concerns resulting from the false notion that a harmful reaction in rats is necessarily indicative of a similar response in humans. The artificial sweetener saccharin, which is commonly used in diet drinks and food additives, is a well-known example. After research found bladder tumors in rats after treatment with large doses of saccharin in the 1970s, there was a lot of public concern. The bladder toxicity was discovered to be caused by a rat-specific protein called -2μ-globulin that has no relevance to humans after mechanistic toxicologists explored the molecular mechanisms underlying this reaction – which only happened in male rats and not females. Lois Leehman-McKeeman and colleagues quickly discovered that this problematic protein was causing false kidney tumor alarms in male rats exposed to other drugs including d-limonene and the gasoline additive methyl tert-butyl ether. This mechanistic research helped alleviate human health worries by linking kidney tumors to a phenomenon known as hyaline droplet nephropathy, which exclusively occurred in male rats. The saccharin panic and subsequent events taught contemporary toxicology the need to fully comprehend the processes behind the hazardous reactions of organisms, both human and nonhuman, to foreign substances. The mechanistic toxicologists who do this work often have excellent backgrounds in biochemistry, genetics, pharmacology, physiology, immunology, and molecular biology, among other basic biomedical sciences. Chemical regulators have recognized the significance of basing their judgments on thorough, empirically proven mechanistic understandings of chemically produced illness in recent decades, and such mechanistic insights have become more important during the risk assessment process.

2.1.4 Regulatory Toxicology

Once the toxicological profile of a new chemical is established, the decision to proceed with its use within a particular sector of the economy is made via a review process that involves applications by sponsoring chemical or pharmaceutical companies to appropriate regulatory bodies. In most jurisdictions, the approval of new chemicals is subject to tight legislative control and usually involves specific regulatory bodies that focus on chemicals used within commercial settings, including pharmaceuticals, agrochemicals, veterinary products, industrial chemicals, etc. Regulatory toxicologists working within these public entities are typically trained in the evaluation of scientific literature and commercial trials which examine the fate and toxicity of chemicals under conditions that approximate their likely use within the 'real world'. Regulatory bodies are also responsible for setting 'safe' exposure levels for chemicals, which are likely to contaminate the human food chain via their use in agriculture and food production. Since analytical chemists can frequently detect very low levels of synthetic chemicals in common food items, it is important to know what levels can be safely consumed.

2.1.5 Clinical Toxicology

Recent decades have seen the emergence of a specialized subgrouping among medical practitioners who are interested in the diagnosis and management of chemically induced diseases. This can include emergency medicine specialists who are experts in the treatment of acutely poisoned individuals such as victims of deliberate or accidental overdosing with drugs or workplace chemicals. Yet there is much more to clinical toxicology than poisoning management alone since the skills of trained clinicians are also essential in the medicolegal setting where it is important to judge if particular symptoms or diseases experienced by a patient are due to work-related chemical exposures or whether they result from pre-existing diseases or other kinds of exposures. Since these cases demand skills that extend beyond the knowledge of drug and chemical toxicity that is acquired in a typical medical training program, medical graduates seeking a career in clinical toxicology usually complete additional specialized graduate programs in pharmacology, toxicology, and environmental health.

2.1.6 Occupational Toxicology

The use of synthetic chemicals within the workplace raises critical issues of special concern to occupational toxicologists. Recognizing the natural duty of care employers bear towards their employees, modern democracies have enacted legislative frameworks that promote the safety and health of workers in specific occupational settings. Workplaces that involve repetitive use of noxious chemicals often receive extra scrutiny. For example, in addition to workers in the chemical industry, personnel within the mining industry and other primary industries are regularly exposed to hazardous substances. Occupational toxicologists are responsible for monitoring chemical hazards within these workplaces while keeping abreast of ongoing toxicological research concerning the health effects of substances relevant to their industrial sector. Their responsibilities can include overseeing the use of personal and environmental monitoring devices to assess chemical concentrations within the workspace of workers as well as ensuring regular health checks to assess health endpoints and biomarkers that are relevant to the chemicals used by each employee. Preparation for a career in occupational toxicology usually requires strong training in toxicology and chemistry together with occupational health and safety, public health, and epidemiology.

2.1.7 Forensic Toxicology

Sometimes relations between workers and employers in particular work settings can deteriorate, leading to legal proceedings in which chemical exposures often take center stage. In such cases, the complainants may be family members of workers who died of a disease that they suspect was work-related. Such situations require the input of forensic toxicologists with appropriate medicolegal and analytical training to assist courts and juries in establishing the validity of such claims. As we saw in the case of Mathieu Orfila in nineteenth-century France, forensic toxicologists also possess expertise concerning the detection of drugs and poisons within body fluids and autopsy tissues obtained from homicide victims and other cases of unexpected death. Such scenarios are often highly complex and subject to ambiguity concerning the effects of concurrent chemical exposures that include prescription medicines and recreational drugs such as alcohol, tobacco, or cannabis. Forensic toxicologists also require training in topics unique to this branch of science, including the study of the fate of chemicals with decaying corpses and their rates of degradation by the microorganisms that cannibalize the bodily remains of deceased individuals. Forensic toxicologists typically require a strong background not just in toxicology, microbiology, and pharmacology but also training in forensic science, legal theory, and court proceedings.

2.1.8 Environmental Toxicology

Environmental toxicology explores the broader environmental impact of chemical exposure, with a particular concern for the effect of chemicals on nonhuman target species including birds, terrestrial organisms, and marine species. This complex field requires attention to the stability of pollutants within soil and water and the ability of chemicals to accumulate within the food chain. Working with ecologists, environmental toxicologists assess any role of agricultural chemicals in the extinction of native species or monitor changes in speciation patterns within the proximity of mines or industrial plants. Environmental toxicologists also work with regulatory toxicologists and environmental engineers to assess the effectiveness of remediation efforts at contaminated land sites. The research tools used by environmental toxicologists are increasingly sophisticated and, in addition to the use of analytical technologies to measure levels of pesticides or herbicides in soil, air, or water samples, now include the use of genomic fingerprinting to monitor the effect of agrochemicals and pollutants on populations of microflora and microfauna in soil samples collected from contaminated sites. In addition to training in general toxicology, a background in molecular biology, molecular genetics, ecology, environmental law, and population biology provides a solid foundation for a career in environmental toxicology.

2.1.9 Food Toxicology

Food toxicology is involved in delivering safe and edible food to the consumer. During processing, several substances may be added to food to make it look, taste, or smell better. Fats, oils, sugars, starches, and other substances may be added to change the texture and taste of food. All these additives are studied to determine if and at what amount, they may produce adverse effects. A second area of interest includes food allergies. Almost 30% of the American people have some food allergy. For example, many people have trouble digesting milk and are lactose intolerant. In addition, toxic substances such as pesticides may be applied to a food crop in the field, while lead, arsenic, and cadmium are naturally present in soil and water, and may be absorbed by plants. Toxicologists must determine the acceptable daily intake level for those substances.

2.1.10 Toxicogenomics

Toxicogenomics is a discipline of science that is concerned with the collection, interpretation, and storage of information on gene and protein activity within a particular cell or tissue of an organism in response to xenobiotics. This branch combines toxicology with genomics, as it explains molecular mechanisms involved in the expression of toxicity and derives molecular expression patterns that predict toxicity or the genetic susceptibility to it. In pharmaceutical research, toxicogenomics may be defined as "the study of the structure and function of the genome in response to adverse xenobiotic exposure." This study focuses on inter-individual variations in whole genome or candidate gene single-nucleotide polymorphism maps, haplotype markers, and alterations in gene expression, which may be correlated with drug responses.

2.1.11 Nanotoxicology

Nanotechnology is rapidly becoming a major part of the world economy and our everyday lives. According to some estimates, nanotechnology promises to exceed the impact of the Industrial Revolution. Nowadays, large amounts of nanoscale materials are produced and enter the environment; however, very little is known about their interactions with biological systems. The applications of this novel material can range from structural engineering, electronics, optics, alternative energy, and soil and water remediation to medical uses as therapeutics, diagnostic, or drug delivery devices. Nanoparticle research is currently an area of intense scientific interest, and recent studies indicate that some nanoparticles are toxic. The challenge for toxicologists is to identify key factors that can be used to predict the toxicity of this novel material. However, the characteristics of the nanoparticles present a challenge to toxicological evaluation, since this material in biological matrices may assume different properties than those that occur in bulk samples of the same material. Such behavior often affects nanoparticle composition, dissolution, surface area and characteristics, size, distribution, and shape. All these factors influence the toxicity and largely determine the biochemical and physical impact of the nanomaterial. Therefore, the qualitative and quantitative challenges posed by nanomaterials to toxicology may require the development of new methods and approaches. Understanding nanoscale properties is a critical step for risk assessment and one of the aims of this new branch in toxicological research. Nanotoxicology is also engaged in developing reliable, robust, and data-assured test protocols for nanomaterials genotoxicity testing.

2.2 Importance of Toxicology

The variety of potential adverse effects and diversity of chemicals and living creatures make toxicology a broad discipline of science. It has an impact on human life from the stage of conception in the womb to the last moment of life. There are certain chemicals that can harm even the sperm and ova. Certain other teratogenic chemicals can malform the fetus. The study of toxicology may be useful to human beings in many ways:

1. The data on acute toxicity tests for various chemicals against various organisms may be valuable in the following ways:
    - To have an idea of toxic doses of xenobiotics for certain organisms,
    - Fixation of sublethal doses for long-term toxicity tests,
    - Evaluation of safe doses of those toxicants for certain organisms,
    - Recommendation of maximum permissible limits of those substances in ambient air and drinking water,
    - Evaluation and recommendation of maximum acceptable daily intake,

- Determination of sensitive species,
- Determination of suitable indicator species, and
- Developing air and water quality criteria.

2. The data on long-term toxicity tests may be reliable for:
   - Evaluation of the safe level of chemicals,
   - Development of sensitive predictive tests useful in obtaining information for risk assessment,
   - Rational therapy of intoxications,
   - Determination of the most sensitive stage of a particular organism, and
   - Elucidation of mode of toxicity of substances.
3. The knowledge of toxicology may be helpful in the development of:
   - Suitable drugs against diseases,
   - Suitable and safe food additives, and
   - Suitable and safe pesticides.
4. An understanding of the mechanisms of toxic action of chemicals also contributes to the basic knowledge of physiology, pharmacology, cell biology, and biochemistry.
5. Forensic toxicology may help detect the cause of death after post-mortem investigations in suspected and disputed cases.
6. Analytical toxicology provides suitable methods to evaluate the presence or absence of different substances and their levels in the environment.
7. It may provide information that may be helpful in reducing the level of pollutants and thus in the monitoring of environmental pollution.
8. Toxicology may be helpful in the development of suitable antidotes that are useful in the treatment of cases of occupational, accidental, or intentional poisonings and thus helpful in protecting human beings and other forms of life.

Chapter 3

Navigating Toxicity: Concepts and Principles

"Toxicants: know them to manage them"

3.1 Introduction to Toxicants

Any material that is toxic is referred to as a toxicant. Toxicants are poisonous substances that can be man-made or found naturally. A toxin, on the other hand, is a poison produced naturally by an organism (e.g., plant, animal, insect). Toxicants can be found in various forms in the air, land, water, and food. Toxicant is the preferred catch-all term in the toxicological lexicon for poisonous compounds of both synthetic and natural origin. This term refers to a substance that has the potential to cause cellular and tissue damage under common exposure conditions, whether intentional, unintentional, or both. The term's main drawback is that it does not glide off the tongue as easily as other phrases, which may explain why this scientifically accurate term has not yet made its way into everyday speech – and why inappropriate use of the terms 'toxins' and 'poisons' is still prevalent in scientific literature. Toxicant is particularly useful when combined with a prefix that specifies the site of toxic action for a certain substance; for example, alcohol is a

Environmental Toxicology: Concepts and Principles
Mohd Salim Reshi, Darakhshan Javaid, Syed Sanober Qadri, and Neelu Sinha

ISBN 978-981-5129-68-7 (Hardcover), 978-1-003-63707-3 (eBook)
www.jennystanford.com

hepatotoxicant because it damages the liver in large quantities, but cadmium is a nephrotoxicant because it damages the kidney.

3.2 Occurrence of Toxicants

Toxicants can be found in the air, the soil, the water, and even the food we eat. Toxicants in the environment can harm humans. Toxicants in the environment can be found in fish. Toxins can be found in cigarette smoke. Toxins are also included in e-cigarette aerosol. Toxicants can be found in the emissions of a heat-not-burn tobacco product. Most heavy metals are toxicants. Toxicants can be found in diesel exhaust. Toxicants include pesticides, benzene, and asbestos-like fibers like carbon nanotubes. Phthalates, phenols, sunscreens, pesticides, halogenated flame retardants, perfluoroalkyl coatings, nanoparticles, e-cigarettes, and dietary polyphenols are all possible developmental toxicants.

3.3 Exposure of Toxicants

A simple definition of exposure is an interaction between a chemical and a biological system. It is a situation in which organisms are exposed to certain doses of xenobiotics. A chemical cannot cause harm to a biological system unless a significant amount of the chemical or its biotransformation products reaches the proper spot in the body over a longer duration of time. The physical and chemical features of the chemical, the exposure scenario (that is, the route of exposure, the exposure systems, and the duration of exposure), and the biological system's sensitivity all influence the toxic reaction of the biological system. To characterize the potential hazard of a chemical, it is necessary to have knowledge of the type of effect, the dose required to create the impact, the exposure, and the subject.

3.3.1 Chemical Exposure Scenarios

The millions of different molecules in existence, both natural and manmade, along with the rich complexity of living cells, suggest that chemical exposure can result in a wide range of adverse effects. The biological complexity alone ensures that understanding the molecular pathways causing chemical toxicity will be a difficult task. The study of toxic syndromes is further complicated by the nearly infinite possibilities for human exposure patterns: for some chemicals, exposure may be intermittent, occurring only once or twice over a period of seconds, minutes, or hours, while for others, exposure may occur continuously throughout a human lifetime.

Individuals in the latter 'high exposure populations,' who endure heavy, frequent exposure to harmful compounds beyond levels encountered by most people, must be considered in the risk assessment process. Spray painters who work with volatile solvents every day for most of their careers have quite different exposure patterns to hobby mechanics who work with them just intermittently in their home workshop. These considerations must be considered when designing toxicity testing, as it is critical to determine whether the same chemical causes various toxic effects in different exposure circumstances.

Chemical exposures are frequently classified into one of four groups to make research of toxic phenomena easier. To begin, acute exposure occurs when chemical intoxications last less than 24 hours. These types of exposures occur in hospital emergency rooms, where clinical toxicologists must treat individuals who have inadvertently or intentionally eaten large dosages of medicines or other substances. Clinical manifestations of acute intoxication can include restlessness, irritability, headache, delirium, confusion, vomiting, diarrhea, muscle cramping, convulsions, or even death, depending on the substance and amount ingested. Subacute intoxications are toxicities that appear after repeated exposure to substances over many days or up to one month. These syndromes are also therapeutically relevant since they can involve adverse reactions to drugs that arise early in therapy, especially when patients have never been exposed to the drug before. Subchronic exposures, on the other hand, involve continuous, repetitive exposure to a chemical over a period of 1–3 months. This broad category includes both single doses of a substance (e.g., an antibacterial antibiotic used every day to treat a persistent urinary tract infection) and long-term exposures to chemicals found in our food as food additives or toxins in drinking water. Chronic exposures, which last three months or longer, are like the previous category. These might involve either recurrent discrete dosage or long-term exposure to chemicals found in food, drinking water, or the air. Occupational exposures, such as when print shop workers inhale ink solvents daily for years or even decades, are included in this category. The toxicological evaluation of new drugs and chemicals is substantially complicated by the fact that diverse hazardous responses to the same substance might arise depending on exposure patterns. This guarantees that during routine toxicity testing of compounds in rodent bioassays, numerous research designs that mimic diverse patterns of exposure are included. As a result, some tests are designed to detect acute toxic responses after a limited number of doses, while others need chronic testing and study of impacts on animal health over a lifetime of exposure under controlled laboratory circumstances.

3.4 Localization of Toxicity

The human body is extremely complicated, as it is made up of over 100 different cell types that are arranged into highly ordered three-dimensional tissue and organ structures. The assembly and proper functioning of these systems are made possible by a large network of genetic and metabolic pathways. Because toxicants can have complicated and unpredictable impacts on multiple systems, pinpointing the exact physiological region where toxicity manifests is difficult. Triortho cresyl phosphate (TOCP) harmed nerve cells in the central nervous system, diethylene glycol caused kidney impairment, and short-term usage of thalidomide hurt unborn infants without necessarily affecting their mothers, according to the sketch of toxicological history. This tendency for specific toxicants to produce injury in specific bodily regions raises the question of toxicity localization. Toxicologists generally divide toxic syndromes into two categories: local and systemic.

3.4.1 Local Toxicity

To begin with, some drugs cause local toxicity when they come into contact with the body (Fig. 3.1). Such chemicals usually have significant physicochemical characteristics, such as being strongly acidic or alkaline, which assures that they harm whatever tissue they come into contact with. Body parts that come into direct contact with the outside world, such as the skin, lungs, nasal cavities, and eyes, are the most vulnerable to toxicants that act locally. Skin burns, rashes, contact dermatitis, and blistering can result from exposure to such compounds, especially if they are handled manually during one's daily work. Chemicals with high chemical reactivity and evident local toxicity characteristics, such as vesicant gases like phosgene, mustard gas, and chlorine, have received consideration as chemical warfare agents for obvious reasons.

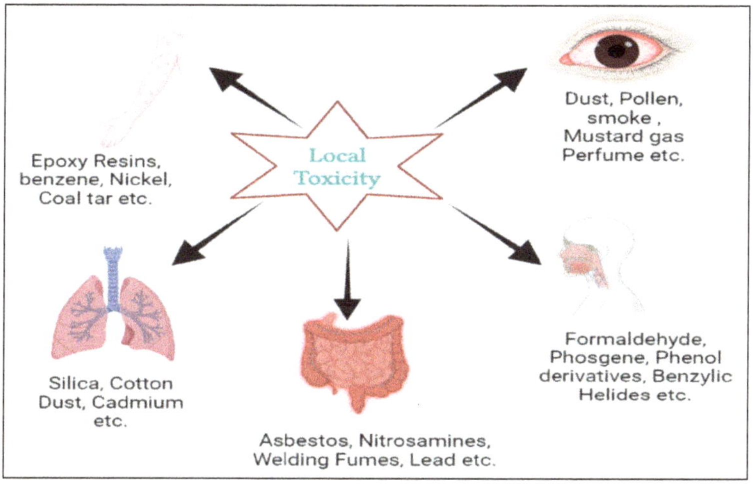

Figure 3.1 Local toxicity: Some chemicals/drugs can result in local toxicity when they come into direct contact with the different parts of the body like skin, eyes, gastrointestinal tract, lungs, and nasal cavity.

3.4.2 Systemic Toxicity (Target Organ Toxicity)

Although local toxicity is important for a small number of substances with unusual properties, most chemicals cause systemic toxicity. Such compounds must be absorbed into the body, where they are dispersed throughout the bloodstream, causing harm to one or more vulnerable organs (Fig. 3.2). Because excessive concentrations of toxicants accumulate in some target organs, they become injured. Eye toxicity from systemic use of the antimalarial drug chloroquine, or life-threatening lung damage from accidental or deliberate intake of paraquat, a commonly used herbicide, are two examples. In both cases, high cellular uptake ultimately results in high chloroquine and paraquat intracellular concentrations within their respective target tissues. Other chemicals affect specific organs that are well-vascularized and perfused with blood, absorbing a large proportion of the toxicant's 'internal' or 'absorbed dose.' In general, a chemical's solubility in nonpolar liquids predicts its absorption properties, as this feature determines its ability to penetrate membranes. Chemicals with lipophilic characteristics are efficiently transported to the liver upon oral consumption, ensuring liver injury is a common result of harmful chemical exposure. In addition to passive diffusion, recent research has revealed that miniature 'pumps' or membrane transporters play a significant role in controlling the transport of foreign chemicals across cell membranes, revealing that chemicals frequently accumulate in target tissues due to high levels of influx pumps. As a result of these variables, the body's main excretory organs, such as the liver and kidney, are frequently exposed to ingested toxicants. Hepatic and renal levels of lead, cadmium, and mercury, for example, are hundreds of times higher than tissue concentrations in other organs because these tissues often accumulate heavy metals prior to their permanent clearance from the body. This ability to sequester such compounds comes at a cost; poisoning with many heavy metals causes liver damage (hepatotoxicity) and kidney injury (nephrotoxicity). Because the tissue expresses large numbers of enzymes that convert the substance to toxic, cell-damaging metabolites, the toxicity of specific compounds frequently localizes to that organ. This mechanism, known as toxicological bioactivation, contributes to a variety of chemically produced clinical conditions. During the 1940s, James and Elizabeth Miller of the University of Wisconsin set the groundwork for our current understanding of the role of bioactivation in chemical toxicity. This husband-and-wife research team discovered that the cancer-causing properties of azo dyes – a class of compounds like those linked to bladder cancer in German textile workers in the 19th century by Rehn – were due to toxic metabolites that attacked DNA and proteins in exposed cells, rather than the parent compound. Target organ toxicity is inherently interesting since it offers a slew of relevant research topics. For example, why does a chemical induce liver damage when a molecule that is closely related targets the kidney? What happens to the chemical within the body – is it metabolized quickly or does it linger in specific tissues? If the latter, are those organs most vulnerable to toxicity, or does toxicity occur elsewhere? What is the dose–response relationship for the toxic syndrome – do the same toxic responses occur at high doses as at low doses, or do different problems emerge due to alternate routes of metabolism that follow saturation of low-capacity pathways? What happens if the same total toxicant dose is administered in small fractionated amounts over an extended timeframe – does toxicity disappear, or do new toxicities emerge? How does the 'route of entry' affect toxic outcomes – is, for example, the substance as toxic following dermal application compared to oral absorption? What are the fundamental mechanisms underlying cellular injury in affected tissues, and why are these actions not exerted on cells in unaffected tissues? How does the body cope with the presence of the chemical if exposure occurs over an extended period – can target tissues increase the levels of specific transporters or enzyme systems to protect themselves? Finally, are individuals from different genetic backgrounds equally vulnerable to the target organ toxicity? Because target organ toxicity raises so many problems, it is no surprise that these symptoms receive more study attention than those caused by chemicals that harm the body's first point of contact.

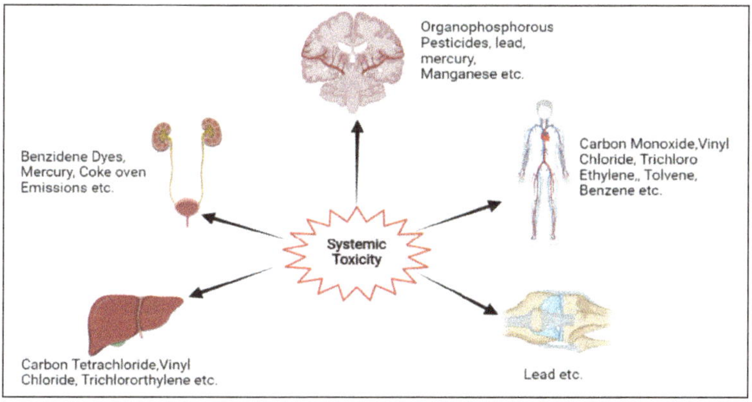

Figure 3.2 Systemic toxicity: Systemic effects caused inside the body once hazardous agents, like pesticides, benzene, mercury, lead, etc., have entered. Systemic effects can occur in the blood and the organs like the brain, kidneys, liver, bones, etc.

3.5 Chemical Interactions

Toxicants in the environment often interact with other chemical and biological factors, which can modify their toxicity. These chemical interactions can be broadly classified into four categories, as illustrated below:

1. Antagonistic Effect: In this type of interaction, one toxicant reduces the effect of another, resulting in a total effect that is less than the sum of the individual effects (1+1 < 2). For example, vitamins E and A can mitigate the response to certain carcinogens. Another instance is the use of calcium gluconate to counteract fluoride poisoning.

2. Additive Effects: When two or more chemicals with similar physiological functions are combined, their effects are additive, meaning the total effect is equal to the sum of their individual effects (1+1 = 2). For example, rats exposed to both lead and arsenic exhibit twice the toxicity compared to exposure to only one of these elements. Similarly, the combined exposure to benzene and toluene results in an additive effect on the central nervous system.

3. Synergistic Effect: In this interaction, the combined effect of two toxicants is greater than the sum of their individual effects (1+1 > 2). For instance, occupational exposure to asbestos can increase the lung cancer rate twentyfold, especially in smokers. Another example is the combined effect of alcohol and barbiturates, which can significantly enhance depressant effects on the central nervous system.

4. Potentiative Effect: This occurs when a non-toxic or less toxic substance enhances the action of a toxicant (1+0 > 1). For example, an increase in pH from 7 to 8 can boost the toxicity of ammonia by 200% in fish. Similarly, the presence of isopropanol can potentiate the hepatotoxic effects of carbon tetrachloride.

3.6 Dose–Response Relationship

The quantity received by an individual is represented in units of mass or molar amount, (e.g., milligrams, micromoles) and is simply referred to as the dose. The dosage (e.g., milligram per kilogram body weight, micromoles per square meter) is often more meaningful when expressed in terms of recipient characteristics (e.g., body weight, body surface area). Paracelsus is credited with being the first to recognize a link between a compound's dose and the response it elicits. Not only does he say that all substances can be hazardous at some doses, but he also says that "the correct dose distinguishes a poison from a medicine," a principle that underpins pharmaceutical therapy. Figure 3.3 depicts a typical dose–response curve, which plots the percentage of organisms or systems that respond to a chemical against the dose. There will be a dose below which no impact or response is observed for many chemicals and effects. This is referred to as the "threshold dose." This notion is important because it indicates that a no observed effect level (NOEL) can be calculated. This value can be used to evaluate the safe intake of food additives and pollutants like pesticides. Although this is commonly acknowledged for most types of chemicals and harmful effects, the form of the curve for chemical carcinogens operating through a genotoxic mechanism is controversial, and their effect is assumed to be a no-threshold phenomenon for regulatory purposes.

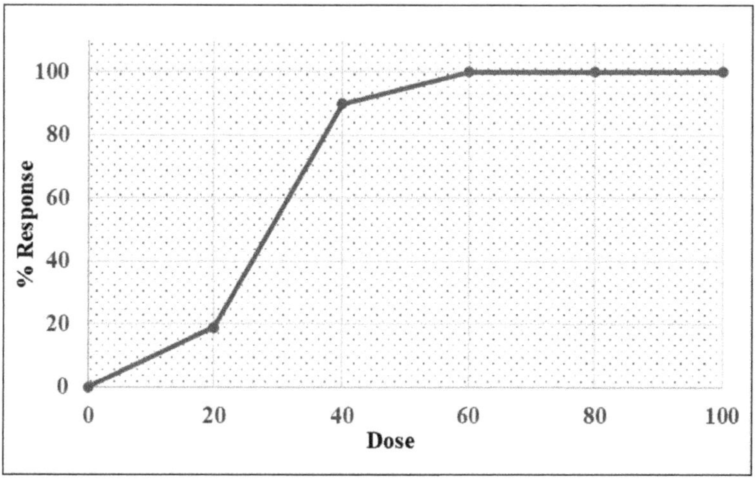

Figure 3.3 A typical dose–response curve (Modified from Burcham, 2013).

3.6.1 Measurement of Dose–Response Relationship

The graded response displayed in Panel A of (Fig. 3.3) depicts the link between the administered dosage and the severity of toxicity within an individual. In this hypothetical scenario, a toxicant causes apoptotic cell death in circulating blood cells, with a higher proportion of cells dying as the toxicant dose is increased (Fig. 3.4).

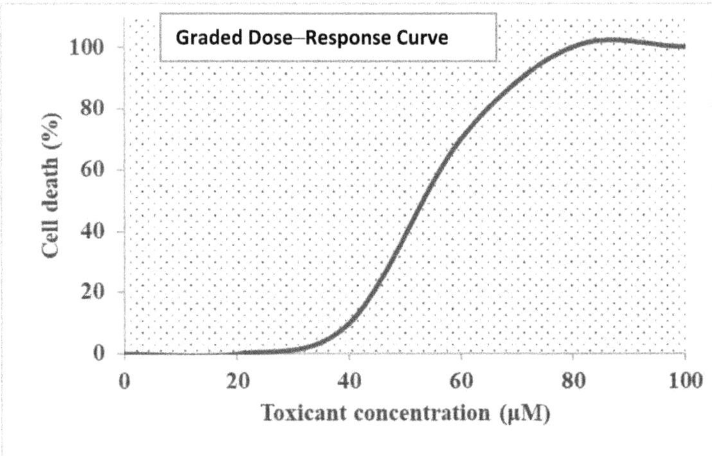

Figure 3.4 Graded dose–response curve (Modified from Burcham, 2013).

The graph implies that an individual's response to a toxicant is graded over a range of dosages, with toxicity severity increasing as doses increase. Such representations reflect the S-shaped sigmoidal curve long known in classic pharmacology, assuming a log scale is employed to indicate the dosage on the x-axis. The dose–response relationship is investigated more frequently at the population level, where a graphical depiction of the applied dosage and the prevalence of the toxic response also provides the recognizable sigmoidal curve (Fig. 3.5).

This type of data is referred to as quantal data since it expresses the toxic response in terms of the percentage of people who are affected in an exposed population. A quantal dose–response curve provides useful quantitative estimates that provide insight into a compound's toxicity. For example, the dose eliciting the reported toxic response in 50% of the population (TD_{50}) and the threshold dose at which toxicity is first observed can both be easily determined

(Fig. 3.5). The no adverse effect level (NOAEL) is a related concept that can be calculated using dose-response data of this type. It is important to remember that these quantitative estimations are not absolute figures chiseled in stone, because the actual number computed for the threshold dose and TD_{50} is dependent on the number of individuals within the population that is under investigation. Research conducted in a population of 20,000 mice to estimate the TD_{50} for a certain food additive will almost certainly give a different result than one obtained from just 10 exposed mice.

Another aspect of quantal representations of toxicological phenomena is that differences in the dose dependence of various toxic consequences are frequently observed. In the hypothetical example depicted in (Fig. 3.6), skin rashes are a very sensitive sign of toxicant intoxication, but hepatotoxicity occurs at greater concentrations, and even higher doses cause CNS depression.

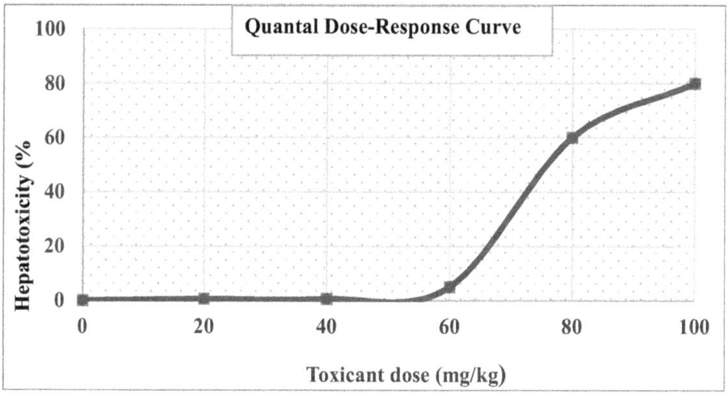

Figure 3.5 Quantal dose–response curve (Modified from Burcham, 2013).

The shape and slope of the dose-response relationships offer important comparisons between different toxicants (Fig. 3.7), just as they do when interpreting dose-response curves in pharmacology. These curves reflect a toxicant's efficacy, or how well it causes the toxic reaction of concern, as well as its potency, or the doses required to elicit its typical toxic effects. Several toxicants are compared for their effectiveness in inducing bladder cancer in a population of laboratory rats in the hypothetical case presented in (Fig. 3.7). Toxicant 'a' is claimed to have stronger carcinogenic potential than toxicants 'b' and 'c' since it causes the maximum production of tumors at lower doses (e.g., 0.1 g/kg). Toxicant 'c', on the other hand, has lower tumorigenic efficacy than toxicants 'a' and 'b' because it induces a lower proportional production of tumors at high dosages. Toxicants 'b' and 'c' have similar potencies.

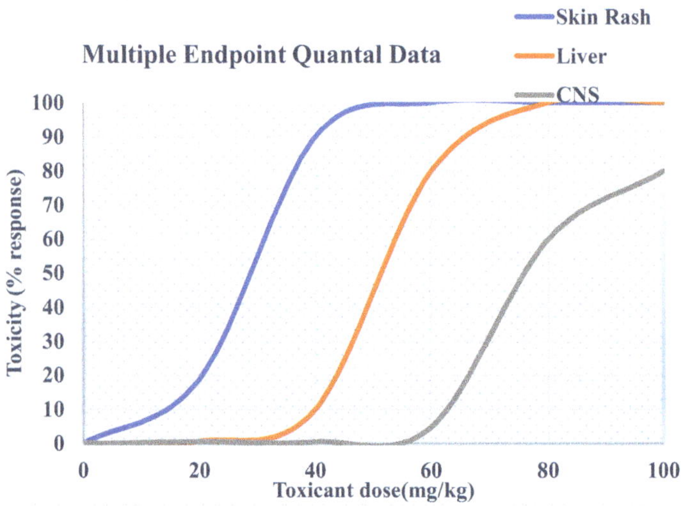

Figure 3.6 Multiple endpoint quantal data (Modified from Burcham, 2013).

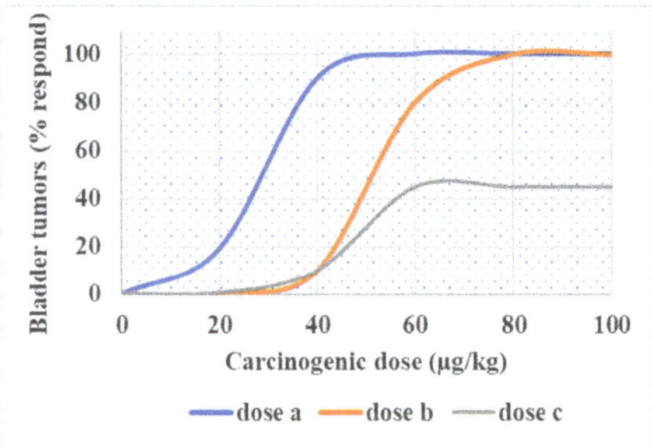

Figure 3.7 Comparison of carcinogenic potency and efficacy of three hypothetical toxicants: 'a', 'b', and 'c'. Toxicant 'a' has the highest potency since it increases tumor incidence most strongly at the lowest doses tested. The fact that toxicants 'a' and 'b' induced a maximal tumor response at high doses indicates they possess comparable efficacy. Toxicant 'c' has comparable potency to toxicant 'b', yet lower efficacy (Modified from Burcham, 2013).

Before we wrap off our brief look at dose–response data, it is important to stress the necessity of establishing this link whenever researchers try to link individual chemicals to specific pathological syndromes or disease outcomes in human populations. Most humans continuously absorb various chemicals through our dietary habits, use of personal care products, consumption of alcohol or tobacco, ingestion of medicines or alternative therapies, or even chemical handling in the home workshop or during other 'after-hours' hobby activities, so this can be a challenging task. It can be difficult to tease out the participation of individual workplace toxicants against the backdrop of normal xenobiotic exposure. When the dose received by each member of the research population is quantified, the strength of disease associations for a given toxicant is increased. Wearing personal exposure monitoring equipment, which assesses the amount of chemicals to which a person is exposed during a working day, is one example. Alternatively, the internal dose of a substance received by each study participant can be determined by assessing the levels of particular compounds and their metabolites in blood or urine samples collected from exposed shift workers. The quality of human risk assessment efforts is considerably enhanced by determining the dose–response relationships for toxic responses of concern for a given xenobiotic. In response to this concern, current toxicology focuses on characterizing worker exposure profiles in specific occupational environments, including the use of personal monitoring devices that measure toxicant amounts in individual employees' workspaces throughout a workday shift.

3.7 Movement of Toxicants in the Environment

Chemicals released into the environment are rarely found in the same form or at the same location where they were released. Agricultural chemicals employed as sprays may drift into the atmosphere as air contaminants or runoff water as water contaminants. Many of these chemicals are susceptible to fungal or bacterial breakdown and detoxification and are frequently broken down into products that can enter the carbon, nitrogen, or oxygen cycles. Other agricultural chemicals, particularly halogenated organic compounds, are resistant to microorganism metabolism to varying degrees and persist as

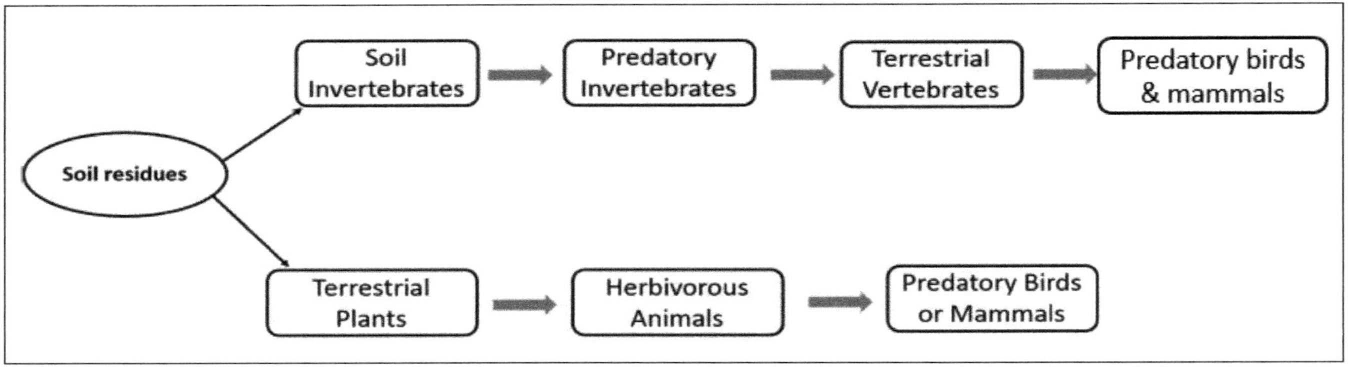

Figure 3.8 Examples of simplified food chains.

contaminants in soil and water; they may enter biological food chains and move to higher trophic levels, or they may persist as food contaminants in processed crops. This scenario can be applied to any toxicant discharged into the environment for a specific purpose or because of industrial processes, combustion, or other means. Chemicals released into the environment are also susceptible to chemical breakdown, which is frequently accelerated by ultraviolet radiation. Although most transport occurs between inanimate phases of the environment, dilution of the toxicant in question and transmission among living creatures can result in increasing concentration or bioaccumulation. Organisms quickly absorb lipid-soluble toxicants after being exposed to them in the air, water, or soil. They remain in the tissues long enough to be transmitted to the next trophic level unless quickly metabolized. At each stage, the lipophilic toxicant is kept while most of the meal is digested, consumed, and eliminated, resulting in an increase in toxicant concentration. The toxicant can become harmful at any point along the chain, especially if the organism at that level is more susceptible than those at the level preceding it. As a result, eggshell thinning in certain raptorial birds was almost certainly caused by DDT and DDE ingestion, as well as their susceptibility to this type of toxicity. Simplified food chains are shown in (Fig. 3.8).

It is obvious that toxicants can be transported across both aquatic and terrestrial food chains, while in the former higher members of the chains, such as fish, can absorb substantial levels of toxicants directly from the medium. The huge area of gill filaments, their close contact with the water, and the high flow rate of water over them cause accumulation. Given these characteristics, as well as a toxicant with a high partition coefficient between lipid membranes and water, significant absorption is inevitable.

Chapter 4

Xenobiotics and Their Biotransformation/Metabolism

"Detoxifying the unfamiliar"

4.1 Introduction

As there is a very fine line between pleasure and pain. They are the two sides of the same coin and one cannot exist without the other. In reality, there is nothing in nature that is solely advantageous to the biological system. Everything a body comes into contact with will have both helpful and detrimental effects, and xenobiotics are no exception to this general rule. The term "xenobiotic" refers to agents that are foreign to the body or a biological system. Xenobiotics are defined as chemicals to which an organism is exposed that are extrinsic to the normal metabolism of that organism. The word Xenobiotic was coined by Dr. Howard Mason in the 1960s. While there has been extensive research, both earlier and ongoing in the field of xenobiotics, it is pertinent to review this field from the clinician's perspective. Translational research in the field of xenobiotics should enrich us with knowledge of the prevention, detection, and treatment of xenotoxicity. Xenobiotics affect the human body such as through inhalation (respiratory tract), ingestion (gastrointestinal tract), dermal contact (skin), and parenteral (circulation/muscles)

Environmental Toxicology: Concepts and Principles
Mohd Salim Reshi, Darakhshan Javaid, Syed Sanober Qadri, and Neelu Sinha

ISBN 978-981-5129-68-7 (Hardcover), 978-1-003-63707-3 (eBook)
www.jennystanford.com

because of intentional or accidental exposure. The potential effect of xenobiotics on humans by causing environmental pollution, organ toxicity, and carcinogenesis are matters of clinical concern. Although an extensive and comprehensive review on the topic of xenobiotics will certainly benefit every researcher immensely, the present article focuses on pertinent aspects from a clinician's viewpoint. A xenobiotic is all pervasive and can originate from the environment, air, water, food additives, drugs, dietary supplements, antioxidants or dyes, emulsifiers, cosmetics, soaps, perfumes, nanomaterials in earth's crust like silica, asbestos, industrial chemicals, or pesticides.

4.2 History

Studies on how people transform the substances that they ingest began in the mid-19th century, with chemists discovering that organic chemicals such as benzaldehyde could be oxidized and conjugated to amino acids in the human body. During the remainder of the 19th century, several other basic detoxification reactions were discovered, such as methylation, acetylation, and sulfonation.

In the early 20th century, work moved on to the investigation of the enzymes and pathways that were responsible for the production of these metabolites. This field became defined as a separate area of study with the publication by Richard Williams of the book *Detoxication Mechanisms* in 1947. This modern biochemical research resulted in the identification of glutathione S-transferases in 1961, followed by the discovery of cytochrome P450s in 1962, and the realization of their central role in xenobiotic metabolism in 1963.

4.3 Classification of Xenobiotics

Xenobiotics are chemical substances that are foreign to a biological system. These substances can be classified based on their nature into several categories:

1. Organic Xenobiotics:

- **Synthetic Organic Compounds**: These include a wide range of man-made chemicals such as pesticides (e.g., DDT, glyphosate), industrial chemicals (e.g., PCBs, dioxins),

pharmaceuticals (e.g., antibiotics, analgesics), and plastics (e.g., BPA, phthalates).
  - **Natural Organic Compounds**: These are naturally occurring substances that may be foreign to certain organisms. Examples include plant alkaloids, mycotoxins, and certain natural food additives.

2. **Inorganic Xenobiotics**:
   - **Metals and Metalloids**: These include heavy metals such as lead, mercury, cadmium, and arsenic. These elements can be toxic in various biological systems, causing a range of adverse effects.
   - **Non-Metallic Elements and Compounds**: These include substances like fluoride, nitrate, and perchlorate, which can also have detrimental effects on biological systems.
3. **Radiological Xenobiotics**:
   - **Radioactive Isotopes**: These include isotopes such as iodine-131, cesium-137, and uranium. They are often byproducts of nuclear reactions and can cause radiation poisoning and long-term health effects such as cancer.
4. **Biological Xenobiotics**:
   - **Pathogenic Organisms and Toxins**: These include bacteria, viruses, fungi, and their toxins (e.g., botulinum toxin, aflatoxins). While they are biological in origin, they are foreign to the host organism and can cause diseases.
5. **Physical Xenobiotics**:
   - **Nanomaterials and Microplastics**: These are emerging contaminants that include nanoparticles and micro-sized plastic particles. Due to their small size and unique properties, they can interact with biological systems in unforeseen ways.

4.4 Transport System of Xenobiotics

4.4.1 Passive Diffusion

Simple passive diffusion is defined as the movement of toxicants from areas of higher concentration to areas of lower concentration, which

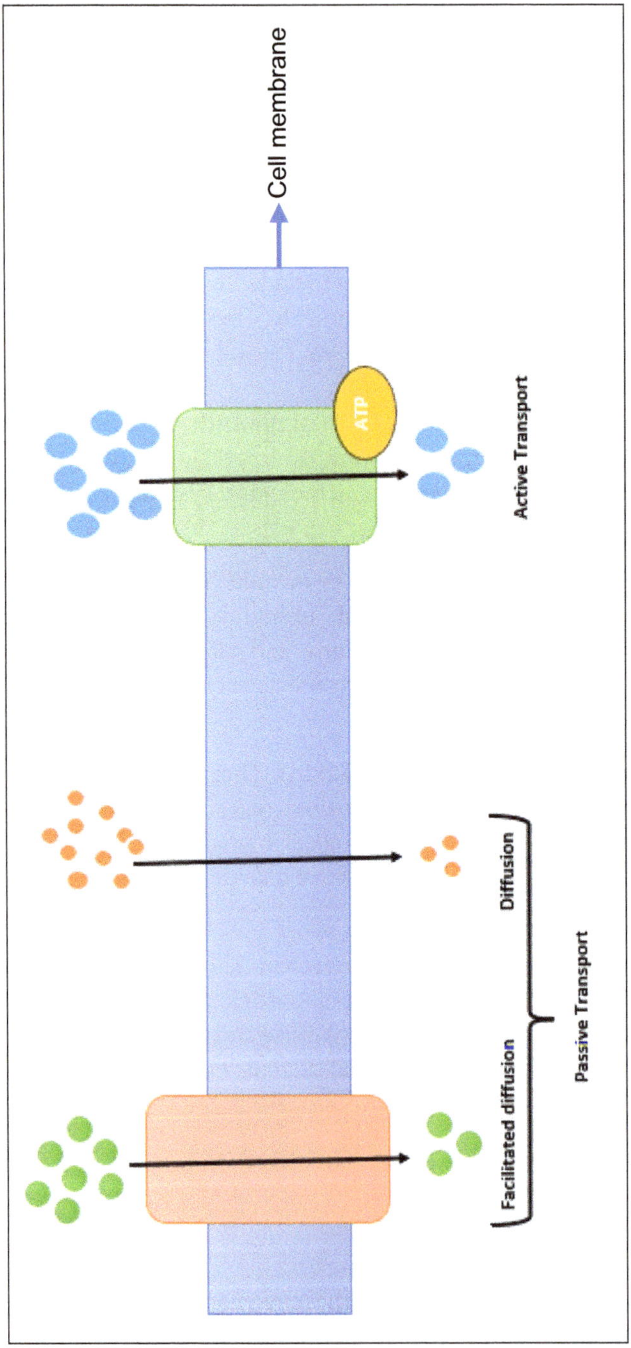

Figure 4.1 Transport of xenobiotics through active and passive diffusion.

does not require energy. On the other hand, when molecules travel from a region of low concentration to a region of high concentration, it is called active transport and requires energy. Most toxicants cross cell membranes by simple and passive diffusion (Fig. 4.1). The rate of passage is related directly to the concentration gradient across the membrane and the lipid solubility. For example, mannitol is hardly absorbed (<2%), acetylsalicylic acid is well absorbed (21%), and thiopental is even more readily absorbed (67%). It is noteworthy that the chloroform and water partition of the non-ionized forms of these chemicals is, respectively, <0.002, 2, and 100. Many toxicants are ionizable. The ionized form is often unable to penetrate the cell membrane because of its low lipid solubility. On the other hand, the non-ionized form is likely sufficiently lipid-soluble, and its rate of penetration depends upon the lipid solubility. The extent of ionization of weak organic acids and bases depends upon the pH of the medium. Thus, for the former, such as benzoic acid, diffusion is facilitated in an acidic environment, where this chemical exists mainly in the non-ionized form; for the latter, such as aniline, diffusion is facilitated in a basic environment.

4.4.2 Filtration

The membranes of the capillaries and glomeruli have relatively large pores (about 70 nm) and allow molecules smaller than albumin (molecular weight 60,000 Da) to pass through. The bulk flow of water through these pores results from hydrostatic and/or osmotic pressure and acts as a carrier of toxicants. The pores in most cells, however, are relatively small (about 4 nm) and allow chemicals only up to a molecular weight of 100–200 Da to pass through. Chemicals of larger molecules, therefore, filter into and out of the capillaries. They can, therefore, establish equilibrium between the concentrations in plasma and the extracellular fluid, but cannot do so by filtration between the extracellular and intracellular fluids.

4.4.3 Carrier-Mediated Transport

This involves the formation of a complex between the chemical and a macromolecular carrier on one side of the membrane. The complex then diffuses to the other side of the membrane, where the chemical is released. Thereafter, the carrier returns to the original surface to repeat the transport process. The carrier has a limited capacity. When it is saturated, the rate of transport is no longer dependent on the concentration of the chemical and assumes zero-order kinetics. Structure, conformation, size, and charge are important in determining the affinity of a chemical for a carrier site, and competitive inhibition occurs among chemicals with similar characteristics. Active transport involves a carrier that moves molecules across a membrane against a concentration gradient, or, if the molecule is an ion, against an electrochemical gradient. It is selective for specific structural features of chemicals. It requires the expenditure of metabolic energy by hydrolysis of adenosine triphosphate and can be inhibited by poisons that interfere with cell metabolism. Facilitated diffusion is like active transport but does not move molecules against a concentration or electrochemical gradient. Further, it is not energy-dependent, and metabolic poisons do not inhibit this process. Therefore, facilitated diffusion is classified as a passive transport system. Glucose, sodium ions, and chloride ions are examples of substances using facilitated diffusion to efficiently cross plasma membranes.

4.4.4 Engulfing by the Cell (Endocytosis)

Particles may be engulfed by cells. When the particles are solid, the process is called phagocytosis and when they are liquid, it is called pinocytosis. This process is also called active transport and requires energy. Such special transport systems are important for the removal of particulate matter from the alveoli and of certain toxic substances from the blood by the reticuloendothelial system. By this process, carrageenans (molecular weight about 10,000–40,000 Da) are also absorbed from the gut.

4.5 Absorption Pathways

The main routes by which toxicants are absorbed are the gastrointestinal tract (GI tract), lungs, and skin. However, in toxicologic studies, special routes such as intravenous, intraperitoneal, intradermal, intramuscular, and subcutaneous injections are also used. Other minor routes include intracardiac and intra-articular injection and intraosseous infusion.

4.5.1 Gastrointestinal Tract

Many toxicants enter the GI tract along with food and water, or alone as drugs or other types of chemicals. Apart from those that are caustic or irritating to the mucosa, most toxicants do not exert any toxic effect unless they are absorbed. Absorption takes place along the entire GI tract. For example, certain drugs are administered as sublingual tablets and suppositories to be absorbed in the mouth and rectum, respectively. However, the mouth and rectum are, in general, insignificant sites for the absorption of environmental chemicals. The stomach is a significant site of absorption, especially for weak acids, which exist in the diffusible, non-ionized, and lipid-soluble form. On the other hand, weak bases are highly ionized in the acidic gastric juice (pH about 2) and therefore not readily absorbable. The difference in absorption is further amplified by the circulating plasma. Weak acids exist mainly in ionized form in plasma and are carried away; whereas, weak bases exist in non-ionized form and diffuse back to the stomach as shown in (Fig. 4.2).

In the intestine, weak acids will exist mainly in the ionized form and hence are less readily absorbable. However, upon entering the blood, they become ionized and thus will not readily diffuse back. On the other hand, weak bases will exist mainly in the non-ionized form; hence, more readily absorbable. However, the Henderson–Hasselbach equation is not the only determination of GI absorption. It should be noted that intestinal absorption is further enhanced by long contact time, the large surface area provided by the villi and microvilli, and the blood flow rate. In the intestine, there are special carrier-mediated transport systems that are responsible for the absorption of nutrients, such as monosaccharides, amino acids, and elements such as iron, calcium, and sodium. However, a few toxicants, such as 5-fluorouracil (5-FU), thallium, and lead, are known to be absorbable from the intestine by active transport systems. Those toxicants have shown potential competition with nutrients because of transportation using the same specialized transporter: 5-FU for the Pyridium transport system, thallium for the iron transport system, and lead for the calcium transporter. In addition, particulate matter, such as those of azo dyes and polystyrene latex, enter the intestinal cell by pinocytosis.

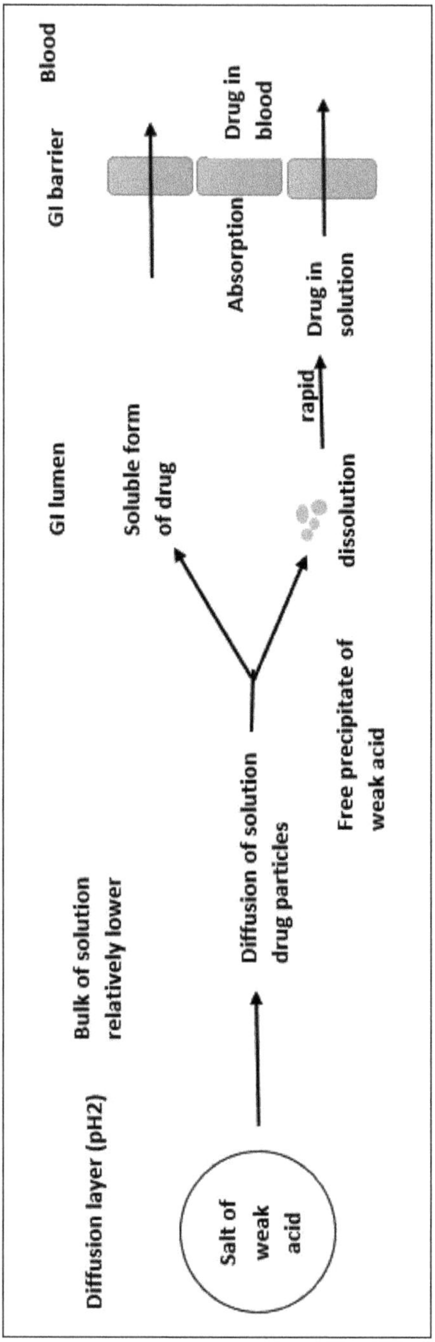

Figure 4.2 Absorption of toxicants by the gastrointestinal tract.

4.5.2 Respiratory Tract

The main site of absorption in the respiratory tract is the alveoli in the lungs. This is especially true for gases such as carbon monoxide, nitrogen oxides, and sulfur dioxide, and for vapors of volatile liquids such as benzene and carbon tetrachloride. Their ready absorption is related to the large alveolar area, high blood flow, and proximity of the blood to the alveolar air. The rate of absorption is dependent upon the solubility of the gas in the blood—the more soluble it is, the faster the absorption. However, equilibrium between the air and blood is reached more slowly for more soluble chemicals, such as chloroform, compared with less soluble chemicals, such as ethylene. This is because the more soluble a chemical, the more it can be dissolved in the blood. As the alveolar air carries only a limited amount of the chemical, greater respiratory rates and a longer time are required to attain equilibrium. This takes even longer if the chemical is also deposited in fat tissue. In addition to gases and vapors, liquid aerosols, and airborne particles may also be absorbed. The inhalational absorption of gases and vapors is influenced by the partitioning of the substance between the blood and gas phase according to its solubility and tissue affinity. In contrast, the inhalational absorption of aerosols and particles is determined by aerosol or particle size, and the water solubility of any substance contained in the aerosol or particle. In general, large particles (>10 μm) do not enter the respiratory tract; when they do, they are deposited in the nose and disposed of by wiping, blowing, and sneezing. Very small-sized particles (<0.01 μm) are likely to be exhaled. Those in the range of 0.01–10 μm are deposited in various parts of the respiratory tract. The larger ones are likely to be deposited in the nasopharynx and absorbed either through the epithelium of this region or through the GI tract epithelium after they are swallowed along with the mucus. Smaller particles are deposited in the trachea, bronchi, and bronchiole, and are then either aspirated onto the mucociliary escalator or engulfed by phagocytes. The particles carried up by the escalator will be coughed up or swallowed. The phagocytes with engulfed particles are absorbed into the lymphatics. Some free particles also migrate into the lymphatics. Soluble particles may be absorbed through the epithelium into the blood. The Task Group on Lung Dynamics (1966) provides a detailed examination of how particles of various sizes are deposited, in different parts of the respiratory tract. However, as a rough estimate, 25% of inhaled particles are exhaled, 50% are deposited in the upper respiratory tract, and 25% are deposited in the lower respiratory tract (Morrow et al., 1967).

4.5.3 Skin

In general, the skin is the largest organ and relatively impermeable, and therefore, it constitutes a good barrier, separating the organism from its environment. However, some chemicals are absorbed through the skin in sufficient quantities to produce systemic effects. A chemical may be absorbed via the hair follicles or through the cells of the sweat glands or sebaceous glands. These are, however, minor routes for absorption because they constitute only a small surface area of the skin. Therefore, percutaneous absorption of a chemical is essentially through the skin proper, which consists of the epidermis and dermis (Fig. 4.3). The first phase of percutaneous absorption is the diffusion of the toxicant through the epidermis, the most important barrier, and especially its stratum corneum. The stratum corneum consists of several layers of thin, cohesive, dead cells that contain a chemically resistant material (protein filament). Small amounts of polar substances appear to diffuse through the outer surface of the protein filaments of the hydrated stratum corneum, and non-polar substances dissolve in and diffuse through the lipid matrix between the protein filaments. All toxicants pass through the stratum corneum by passive diffusion and this process is the rate-limiting step in skin absorption. In the human stratum corneum, there are significant differences in structure and chemistry from one region of the body to another, reflected in the permeability to chemicals. For example, toxicants cross the scrotum readily, cross the abdominal skin less readily, and cross the sole and palm with great difficulty (Zbinden, 1976). The second phase of percutaneous absorption is the diffusion of the toxicant through the dermis, which contains a porous, non-selective, aqueous diffusion medium. Therefore, it is less effective as a barrier than the stratum corneum. Therefore, abrasion or removal of the latter causes a marked increase in the percutaneous absorption. Acids, alkalis, and mustard gases also increase absorption by injuring this barrier. Some solvents, notably dimethyl sulfoxide, also enhance dermal permeability.

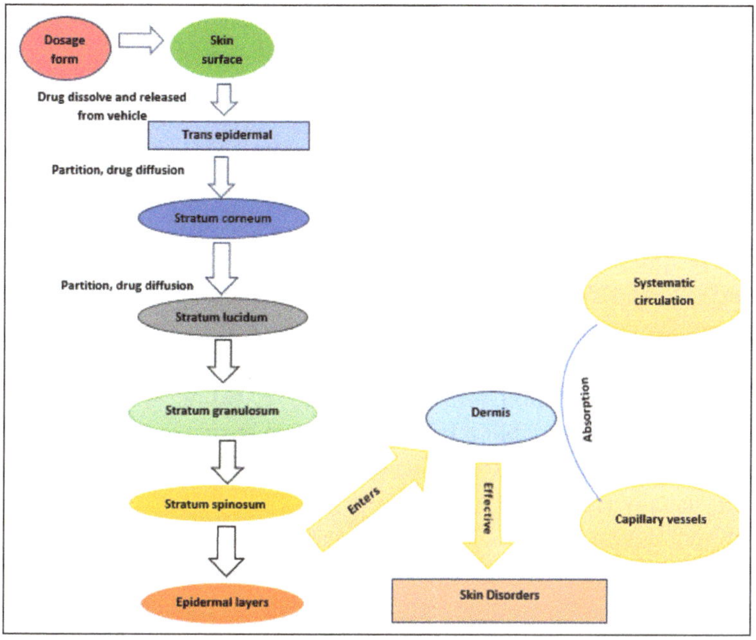

Figure 4.3 The absorption of toxicants through different layers of skin.

4.6 Permeability Barriers and Detoxification

The exact compounds an organism is exposed to will be largely unpredictable and may differ widely over time. These are major characteristics of xenobiotic toxic stress. The major challenge faced by xenobiotic detoxification systems is that they must be able to remove the almost limitless number of xenobiotic compounds from the complex mixture of chemicals involved in normal metabolism. The solution that has evolved to address this problem is an elegant combination of physical barriers and low-specificity enzymatic systems.

All organisms use cell membranes as hydrophobic permeability barriers to control access to their internal environment. Polar compounds cannot diffuse across these cell membranes, and the uptake of useful molecules is mediated through transport proteins that specifically select substrates from the extracellular mixture. This selective uptake means that most hydrophilic molecules cannot enter cells, since they are not recognized by specific transporters. In contrast, the diffusion of hydrophobic compounds across these barriers cannot be controlled, and organisms, therefore, cannot exclude lipid-soluble xenobiotics using membrane barriers.

However, the existence of a permeability barrier means that organisms were able to evolve detoxification systems that exploit the hydrophobicity common to membrane-permeable xenobiotics. These systems, therefore, solve the specificity problem by possessing such broad substrate specificities that they metabolize almost any non-polar compound. Useful metabolites are excluded since they are polar, and in general, contain one or more charged groups.

The detoxification of the reactive by-products of normal metabolism cannot be achieved by the systems outlined above, because these species are derived from normal cellular constituents and usually share their polar characteristics. However, since these compounds are few, specific enzymes can recognize and remove them. Examples of these specific detoxification systems are the glyoxalase system, which removes the reactive aldehyde methylglyoxal, and the various antioxidant systems that eliminate reactive oxygen species.

4.7 Biotransformation and the Enzymes

Once a toxicant is absorbed into an organism through various routes, such as the mouth, blood, lung, skin, eye, or nasal mucosa, it is distributed to various parts of the body, including the excretory organs. Many chemicals are known to undergo biotransformation (metabolic transformation or metabolism), which is catalyzed by enzymes present in the blood and tissues. Although the most important site of such reactions is the liver, other organs also contribute to some extent. Examples are the lungs, stomach, intestine, skin, and kidneys. During the process of biotransformation, "xenobiotics" (foreign substances to an organism or toxic substances) can be either activated to toxic form(s) or inactivated to non-toxic form(s) to become reactive or non-reactive with biomolecules such as DNA, RNA, proteins, and lipids.

There are two types of biotransformation:

1. **Phase I**: It involves oxidation, reduction, and hydrolysis, which may be considered degradation reactions.

2. **Phase II**: It involves the production of a compound (a conjugate) that is biosynthesized from the toxicant, or its metabolite, plus an endogenous substrate may be considered a conjugation reaction.

Biotransformation is, therefore, a process that, in general, converts the parent compounds into metabolites and then forms conjugates, but it may involve only one of these reactions. In other words, it is defined as a converting process of lipophilicity to hydrophilicity, or a process of an absorbable to an excreted form. For example, benzene undergoes oxidation, a phase I reaction, to form phenol, which conjugates with sulfate, a phase II reaction. However, when the administered chemical is phenol, it will be conjugated with sulfate without a phase I reaction. The metabolites and conjugates are usually more water-soluble and more polar, hence more readily excretable. Biotransformation mediated by phases I and II metabolizing enzymes can, therefore, be considered a mechanism of detoxication or bioinactivation, in general, by the host organism. However, it must be noted that in certain cases metabolites are more toxic than parent compounds. Such reactions are known as "bioactivation." This is frequently noted with chemicals that induce cancer where the metabolite and not the parent compound is the culprit. In fact, most carcinogens require metabolic activation by certain enzymes to induce cancer. Aflatoxin B1 and benzo(a)pyrene are such examples. The rate of biotransformation and the type of biotransformation of a toxicant often differ from one species of animal to another and even from one strain to another, a fact that often accounts for differences in toxicity in these animals. The age and gender of the animal and its exposure to other chemicals may also alter biotransformation. Genetic polymorphisms in xenobiotic-metabolizing enzymes and transporters have been considered a critical factor for individual susceptibility or variation in toxic responses (Johansson and Ingelman-Sundberg, 2011). Knowledge of such factors is important in the design of toxicological studies and the interpretation of health hazards attributed to human toxicants.

4.7.1 Phase I (Degradation) Reactions

The three types of phase I reactions, namely, oxidation, reduction, and hydrolysis, are briefly described. Although many exceptions exist, as a general rule, the enzymes show broad substrate specificities because a limited number of enzymes metabolize a great number of substrates.

4.7.1.1 Oxidation

The biotransformation of a great variety of chemicals involves oxidative processes. The most important enzyme systems catalyzing these processes include cytochrome P450 (CYP) and NADPH CYP reductase. In these reactions, one atom of molecular oxygen is reduced to water and the other is incorporated into the substrate as follows:

$$\mathbf{SH + O_2 + NADPH + H \rightarrow SOH + H_2O + NADP}$$

where S is the substrate. Cytochrome-linked monooxygenases (oxidases) are located in the endoplasmic reticulum. When a cell is homogenized, the endoplasmic reticulum breaks down into small vesicles known as microsomes. Because of the location of these enzymes and the wide variety of chemicals catalyzed, they are also known as microsomal mixed-function oxidases (MFOs). The human CYP MFO system consists of more than 60 isozymes in humans, which metabolize a variety of chemicals as shown in this chapter. In addition, the oxidation of several toxicants is catalyzed by non-microsomal oxidoreductases, which are located in the mitochondrial fraction in the 100,000 g supernatant of tissue homogenates. Oxidation may take place in a variety of reactions, and often more than one metabolite is formed. The following are some examples.

4.7.1.2 Microsomal oxidation

1. Aliphatic oxidation involves oxidation of the aliphatic side chains of aromatic chemicals: for example, n-propylbenzene → 3-phenylpropan- 1-ol, 3-phenylpropan-2-ol, and 3-phenylpropan-3-ol, as well as aliphatic compounds such as n-hexane.
2. Aromatic hydroxylation generally proceeds through an epoxide intermediate: for example, naphthalene → naphthalene-1, 2-epoxide → 1-naphthol + 2-naphthol.
3. Epoxidation: for example, aldrin → dieldrin.

4. Oxidative deamination: for example, amphetamine → phenylacetone.
5. N-Dealkylation: for example, N, N-dimethyl-p-nitrophenyl carbamate → N-methyl-p-nitrophenyl carbamate.
6. O-Dealkylation: for example, p-nitroanisole → p-nitrophenol.
7. S-Dealkylation: for example, 6-methylmercaptopurine → 6-mercaptopurine.
8. N-Oxidation: for example, trimethylamine → trimethylamine oxide.
9. N-Hydroxylation: for example, aniline → phenylhydroxylamine.
10. P-Oxidation: for example, diphenylmethylphosphine → diphenylmethylphosphine oxide.
11. Sulfoxidation: for example, methiocarb → methiocarb sulfone.
12. Desulfuration involves the replacement of S by O: for example, parathion → paraoxon.

4.7.1.3 Non-microsomal oxidations catalyzed by enzymes in mitochondria, cytosol, and nuclei

1. Amine oxidation: Monoamine oxidase is located in mitochondria and diamine oxidase is a cytosolic enzyme. Both are involved in the oxidation of the primary, secondary, and tertiary amines, such as 5-hydroxytryptamine and putrescine into their corresponding aldehydes.
2. Alcohol and aldehyde dehydrogenations are catalyzed, respectively, by alcohol dehydrogenase and aldehyde dehydrogenase: for example, ethanol → acetaldehyde → acetic acid.

4.7.1.4 Reduction

Toxicants may undergo reductions through the function of reductases. These reactions are less active in mammalian tissues but more so in intestinal bacteria, because the distant part of the intestine, where reduction is more favorable than oxidation, lacks oxygen. A notable example is the reduction of prontosil to sulfanilamide, converting an inactive chemical to an effective antibacterial drug.

4.7.1.5 Microsomal reduction

1. Nitro reduction: for example, nitrobenzene → nitrosobenzene → phenyl hydroxylamine → aniline.
2. Azo reduction: for example, azobenzene → aniline.

Non-microsomal reductions occur via the reverse reaction of alcohol dehydrogenases.

4.7.1.6 Hydrolysis

Many toxicants contain ester-type bonds and are subject to hydrolysis. These are essentially esters, amides, and compounds of phosphate. Mammalian tissues, including the plasma, contain a large number of non-specific esterases and amidases, which are involved in hydrolysis. The esterases, usually located in the soluble fraction of the cell, may be broadly categorized into four classes:

1. Arylesterases, which hydrolyze aromatic esters
2. Carboxyl esterases, which hydrolyze aliphatic esters
3. Cholinesterases, which hydrolyze esters in which the alcohol moiety is choline
4. Acetyl esterases, which hydrolyze esters in which the acid moiety is acetic acid

In contrast to esterases, amidases cannot be classified according to substrate specificity. Of equal importance, enzymatic hydrolysis of amides proceeds more slowly than that of esters, probably a result of the lack of substrate specificity. An example is procainamide, a drug used for arrhythmia, which is metabolized slowly. On the other hand, structurally similar procaine, a local anesthetic containing a carboxyl ester, cannot be used for systemic purposes because it is metabolized too quickly.

4.7.2 Phase II (Conjugation) Reactions

Phase II reactions involve several types of endogenous metabolites that, as noted above, may form conjugates with toxicants per se or their metabolites. These conjugates are generally more water-soluble and more readily excretable. This is because the physiological transport mechanisms for endogenous metabolites also recognize the conjugates and facilitate their excretion.

4.7.2.1 Glucuronide formation

This is the most common and important type of conjugation. The enzymes catalyzing this reaction are uridine diphosphate-glucuronosyl transferases (UGTs) and the coenzyme is uridine-5′-diphospho-α-D-glucuronic acid (UDP-GA). This enzyme is located in the endoplasmic reticulum. There are four classes of chemical compounds that are capable of forming conjugates with glucuronic acid: (1) aliphatic or aromatic alcohols, (2) carboxylic acids, (3) sulfhydryl compounds, and (4) amines. Bilirubin, the product from heme metabolism, is solely glucuronidated for excretion. An individual who lacks certain UGTs exhibits hyperbilirubinemia (i.e., Crigler-Najjar syndrome) due to the absence of bilirubin glucuronidation.

4.7.2.2 Sulfate conjugation

This reaction is catalyzed by sulfotransferases. These enzymes are found primarily in the cytosolic fraction of the liver, kidneys, and intestines. The coenzyme is 3-phosphoadenosine-5′-phosphosulfate. The functional groups of the foreign compounds for sulfate transfer are phenols and aliphatic alcohols as well as aromatic amines.

4.7.2.3 Methylation

This reaction is catalyzed by methyl transferases. The coenzyme is S-adenosylmethionine. Methylation is not a major route of biotransformation of toxicants. Further, it does not always increase the water solubility of the methylated products. The enzymes are located in both the mitochondrial and cytosolic fractions.

4.7.2.4 Acetylation

Acetylation involves the transfer of acetyl groups to primary aromatic amines, hydrazines, hydrazides, sulfonamides, and certain primary aliphatic amines. The enzyme and coenzyme involved are, respectively, N-acetyl transferases (NATs) and acetyl coenzyme A. In certain cases, such as isoniazid, acetylation results in a decrease not only in the water solubility of an amine but also in toxicity. NATs show genetic polymorphism. Fast acetylators acetylate isoniazid to its diacetylated metabolite compared to slow acetylators. For this reason, slow acetylators are more sensitive to isoniazid-induced toxicity compared to fast acetylators.

4.7.2.5 Amino acid conjugation

This conjugation is catalyzed by amino acid conjugates and coenzyme A. Aromatic carboxylic acids, arylacetic acids, and aryl-substituted acrylic acids form conjugates with not only α-amino acids, mainly glycine, but also with glutamine in humans (and certain monkeys) and ornithine in birds. Taurine, an example of β-amino acid, forms a conjugate with cholic acid.

4.7.2.6 Glutathione conjugation

This important reaction is affected by glutathione S-transferases and the cofactor glutathione (GSH, a tripeptide; γ-glu-cys-gly). GSH conjugates subsequently undergo enzymatic cleavage and acetylation, forming N-acetylcysteine (mercapturic acid) derivatives of the toxicants, which are readily excreted. Some examples of chemicals include epoxides and aromatic halogens, which conjugate with GSH. In addition, GSH might conjugate unsaturated aliphatic compounds and displace the nitro groups in chemicals. In the process of biotransformation of toxicants, several highly reactive electrophilic metabolites are formed. In particular, the generation of reactive oxygen species (ROS) is believed to be responsible for the cancer produced by diesel exhaust particles. Some of these metabolites react with cellular constituents and produce cell death, induce tumor formation, or affect immune function. The role of GSH is to react with electrophilic metabolites such as ROS and thus prevent their harmful effects on the cells. However, exposure to excessive amounts of such reactive substances might deplete GSH, thereby resulting in marked toxic effects. An example of the decreased levels of GSH by acetaminophen and concomitant increase in covalent binding to macromolecules. Similarly, 3-methylindole is bioactivated mainly in the lungs and, after diminishing GSH concentrations, produces lung damage. Further, certain GSH conjugates may become toxic.

4.7.2.7 Bioactivation

Certain chemically stable compounds may be converted into chemically reactive metabolites. The reactions are generally catalyzed by cytochrome P450-dependent monooxygenase systems, but other enzymes, including those of the intestinal flora, are involved in certain cases. Further, additional phase I or phase II reactions may be required. The reactive metabolites, such as epoxides, become covalently bound to cellular macromolecules and induce necrosis and/or cancer. Others, such as free radicals, produce lipid peroxidation resulting in tissue damage.

4.7.2.8 Epoxide formation

Many aromatic compounds are converted to epoxides by microsomal mixed-function oxidase systems. The biotransformation of bromobenzene to its epoxide and subsequent reactions serves as an interesting example of bioactivation and its consequences. Although bromobenzene epoxide may become covalently bound to tissue macromolecules and produce injury, the alternative routes of metabolism may prevent or reduce the injury. The most important of these routes is conjugation with GSH. This reaction is important in that it serves as a protective mechanism. Only after hepatic levels of reduced GSH have been greatly depleted will the bromobenzene epoxide significantly bind to macromolecules and result in hepatic necrosis. Depletion of GSH occurs when an excessive dose of bromobenzene is present or when there has been the induction of microsomal enzymes; both conditions increase the amount of bromobenzene epoxide. Other reactions include a non-enzymatic arrangement to form p-bromophenol and the formation of 3,4-dihydro-3,4-dihydroxybromobenzene catalyzed by hydrolyzing enzymes.

Other chemicals that undergo epoxidation include aflatoxin B1, benzene, benzo(a)pyrene, furosemide, olefins, polychlorinated, polybrominated biphenyls, trichloroethylene, and vinyl chloride. Bioactivation takes place mainly in the liver, and the resulting reactive metabolites induce toxicity through covalent binding with macromolecules in the tissue, resulting in necrosis or cancer formation.

4.7.2.9 N-hydroxylation

Microsomal enzymes from many tissues might N-hydroxylate a variety of chemicals. Some N-hydroxy metabolites, such as those of acetaminophen, 2-acetylaminofluorene, and certain amino azo dyes, induce cancer or tissue necrosis through covalent binding; whereas, others, such as certain aromatic amines, produce hemolysis or methemoglobinemia. N-hydroxy metabolites are also subject to conjugation reactions. Their conjugates with glucuronic acid are readily excreted, those formed with sulfuric or acetic acid; however, may be unstable; and thus be mutagenic, carcinogenic, and highly toxic. 2-Acetylaminoflurene, safrole, and 7,12-dimethylbenz[a] anthracene are some examples.

4.7.2.10 Free radical and superoxide formation

Free radicals are unstable and highly reactive molecules that are capable of existing independently, contain one or more unpaired electrons, have an extremely short half-life, and seek to gain additional electrons to have complete pairs or to bind to other molecules to increase their stability. Certain halogen-containing compounds undergo metabolism to form free radicals. For example, carbon tetrachloride forms a trichloromethyl radical, which initiates the peroxidation of polyunsaturated lipids and covalently binds to protein and unsaturated lipids. These initial reactions are followed by disturbances of various cellular components. Halothane and bromotrichloromethane are other examples of chemicals that may form free radicals. The herbicide paraquat is known to produce superoxide radicals, which produce pulmonary toxicity. It should be noted that during the metabolic process, some chemicals produce not only their reactive metabolites, but also generate free radicals such as ROS (radical forms: superoxide anion, hydroxyl radical, peroxyl, and alkoxyl radical; non-radical forms: O_3, H_2O_2, hypochlorous acid, and singlet oxygen) or chemical radicals. For example, benzo(a)pyrene (BaP) produces an ultimate metabolite [e.g., benzo(a)pyrene 7,8-diol 9,10-epoxide (BPDE)], ROS, and BaP radical cation. Free radicals such as ROS, reactive nitrogen species (radical forms: nitric oxide and nitrogen dioxide; non-radical forms: peroxynitrite, nitroxyl anion, nitrosyl cation, nitronium anion, nitrous acid, etc.), and organic radicals activate non-reactive toxicants to form biologically reactive intermediates or reactive toxicants without metabolizing enzymes. Free radicals play a central role in toxicological sciences and are involved in various types of toxic manifestations and common mechanisms of toxic actions. Therefore, toxic effects may be ameliorated or prevented by scavenging free radicals with antioxidants, such as antioxidant vitamins, curcumin, isoflavonoids, resveratrol, and plant polysaccharides.

4.8 Other Pathways

Ethanol can be oxidized by a dehydrogenase to acetaldehyde, which has been implicated in some of the manifestations of alcohol toxicity. Pyrrolizidine alkaloids are dehydrogenated to reactive pyrrole derivatives, which are carcinogenic. There is evidence confirming that the acute toxicity of aliphatic nitriles is attributable to the cyanide released through hepatic microsomal enzyme activities (Willhite and Smith, 1981).

4.8.1 Activation in the GI Tract

Nitrites and certain amines react in the acidic environment of the stomach to form nitrosamines, many of which have been shown to be potent carcinogens, and nitrates, which, under certain conditions, are converted to nitrites that may induce methemoglobinemia. The artificial sweetener cyclamate is converted by intestinal bacteria to cyclohexylamine, which induces testicular atrophy. Cycasin is converted by intestinal microbiota to its aglycone, methylazoxymethanol, which is hepatotoxic and produces tumors. The roles of intestinal microbiota in the metabolic activation of certain drugs and toxicants are now evident. Intestinal microbiota increases the toxicity of chemicals by producing toxic or carcinogenic metabolites which sometimes would not be formed in tissues.

4.9 Sites of Metabolism/Transformation of Xenobiotics

Quantitatively, the smooth endoplasmic reticulum of the liver cell is the principal organ of drug metabolism, although every biological tissue has some ability to metabolize drugs. Factors responsible for the liver's contribution to drug metabolism include that it is a large organ, that it is the first organ perfused by chemicals absorbed in the gut, and that there are very high concentrations of most drug-metabolizing enzyme systems relative to other organs. If a drug is taken into the GI tract, where it enters hepatic circulation through the portal vein, it becomes well-metabolized and is said to show the first-pass effect.

Other sites of drug metabolism include epithelial cells of the gastrointestinal tract, lungs, kidneys, and the skin. These sites are usually responsible for localized toxicity reactions.

4.10 Factors Affecting Drug Metabolism

The duration and intensity of pharmacological action of most lipophilic drugs are determined by the rate they are metabolized to inactive products. The Cytochrome P450 monooxygenase system is the most important pathway in this regard. In general, anything that increases the rate of metabolism (e.g., enzyme induction) of a pharmacologically active metabolite will decrease the duration and intensity of the drug action. The opposite is also true (e.g., enzyme inhibition). However, in cases where an enzyme is responsible for metabolizing a pro-drug into a drug, enzyme induction can speed up this conversion and increase drug levels, potentially causing toxicity.

Various physiological and pathological factors can also affect drug metabolism. Physiological factors that can influence drug metabolism include age, individual variation (e.g., pharmacogenetics), enterohepatic circulation, nutrition, intestinal flora, or sex differences. In general, drugs are metabolized more slowly in fetal, neonatal, and elderly humans and animals than in adults.

Genetic variation (polymorphism) accounts for some of the variability in the effect of drugs. With N-acetyltransferases (involved in phase II reactions), individual variation creates a group of people who acetylate slowly (slow acetylators) and those who acetylate quickly, split roughly 50:50 in the population of Canada. This variation may have dramatic consequences, as the slow acetylators are more prone to dose-dependent toxicity.

Cytochrome P450 monooxygenase system enzymes can also vary across individuals, with deficiencies occurring in 1–30% of people, depending on their ethnic background.

Dose, frequency, route of administration, tissue distribution, and protein binding of the drug affect metabolism. Pathological factors can also influence drug metabolism, including liver, kidney, or heart diseases.

The silico modeling and simulation methods allow drug metabolism to be predicted in virtual patient populations prior to performing clinical studies in human subjects. This can be used to identify individuals most at risk from adverse reactions.

Chapter 5

Pesticide Toxicity

"The hidden dangers in our pest control"

5.1 Introduction to Pesticides

Pesticides are described as "any chemical or mixture of compounds designed to prevent, eradicate, repel, or mitigate pests (insects, rodents, nematodes, weeds, fungus, etc.)." Insects pose a serious threat to human health because some of them act as vectors, transmitting bacteria, viruses, and parasites (malaria, filariasis, West Nile virus, Zika virus, yellow fever, viral encephalitis, Rickettsialpox, typhus, bubonic plague, and others) from animals to humans or from humans to humans. Millions of individuals fall ill every year, and a huge proportion of them die because of these devastating diseases. In such cases, the use of insecticides is necessary to prevent and/or battle vector-borne illnesses. In addition to protecting human and animal health, pesticides are used to safeguard food from production to consumption. Although most pesticides used today are synthetic compounds, farmers have utilized a variety of natural agents for thousands of years. Nicotine from tobacco, pyrethrins from *Chrysanthemum cinerariaefolium* flowers, and numerous lead, copper, and arsenic compounds are among these

Environmental Toxicology: Concepts and Principles
Mohd Salim Reshi, Darakhshan Javaid, Syed Sanober Qadri, and Neelu Sinha

ISBN 978-981-5129-68-7 (Hardcover), 978-1-003-63707-3 (eBook)
www.jennystanford.com

chemicals. In addition to insects, weeds are a major annoyance for farmers. Farmers spent a lot of time manually clearing weeds before pesticides were introduced. Pesticides to combat other pests, such as fungi and rodents, have also been produced.

5.2 Adverse Effects of Pesticides

Adverse impacts affect both human health and the environment. Accidental acute poisonings have the most severe impacts on people. Several severe outbreaks of methyl and ethyl mercury compounds, hexachlorobenzene (a fungicide), and methyl and ethyl parathion (an organophosphorus pesticide) poisoning have occurred in various regions of the world, sickening thousands of people and causing hundreds of fatalities. Individual occurrences of acute poisoning have often arisen from the unintentional or intentional absorption of significant doses of pesticides. Workers involved in pesticide manufacture, formulation, and application may be exposed to pesticides at work. Pesticides enter the body mostly through the respiratory tract and dermal absorption, but tiny quantities may also enter the gastrointestinal system via contaminated hands and utensils; this is the most common route of exposure in cases of unintentional ingestion of household items by children. Pesticides that are acutely hazardous are more likely to cause this sort of poisoning. The greatest public health worry, however, is pesticide residue consumption in foods, which can affect huge populations (particularly children) over a prolonged period of time.

In addition to human health hazards, pesticides may have a serious impact on the environment. Apart from large-scale accidental release in the environment, only minimal levels are found in various environmental media. Pesticides that are persistent and/or have a proclivity for biomagnification, on the other hand, are likely to have greater amounts. The concentration of a pesticide increases as it goes up the trophic chain in the latter situation. The bald eagle, for example, was nearly extinct owing to the deadly effect of high amounts of dichlorodiphenyltrichloroethane (DDT) bioaccumulated through the bird's polluted food chain. As a result of biomagnification, excessive quantities of DDT and heptachlor were discovered in Singapore's seafood, putting the inhabitants who eat seafood daily at a much-increased risk of cancer. Because polluted soil and water result in contaminated drinking water and human food production, such environmental pollution may also have an impact on human health.

5.3 Categories of Pesticides

Pesticides are usually classified based on their chemical composition and the type of pests they kill.

5.3.1 Insecticides

This is the largest group of pesticides and consists of different chemical subgroups.

5.3.1.1 Organophosphate insecticides

Organophosphate (OP) insecticides are esters of phosphoric, phosphonic, or thiophosphoric acid. Examples of some commonly used OP insecticides include chlorpyrifos, diazinon, dichlorvos, fenthion, malathion, and parathion, and their chemical structures are shown in Fig. 5.1. Their toxicities vary widely. OPs produce toxicity by inhibiting acetylcholinesterase (AChE) primarily at the synapses and neuromuscular junctions (NMJ), resulting in the accumulation of acetylcholine (ACh). Excess ACh induces a variety of symptoms and clinical signs. The severity of symptoms is correlated with the extent of the inhibition of AChE in blood, but the precise relationship varies depending on the OP compound. Muscarinic ACh receptor (mAChR)-associated effects include hypersalivation, lacrimation, urination, diarrhea, tracheobronchial secretion, vomiting, cyanosis, and miosis. Nicotinic ACh receptor (nAChR)-associated effects include muscle tremors, convulsions, and seizures. Evidence suggests that some OPs directly interact with mAChRs and nAChRs. High-dose OP exposure leads to paralysis and death. In OP-induced excitotoxicity, N-methyl-D-aspartate receptor (NMDAR), gamma-aminobutyric acid receptor (GABAR), and other neurotransmitter systems are also involved. Therefore, antidotal treatment includes timely administration of a mAChR blocker, atropine sulfate (0.5 mg/kg; one-third IV and the remainder IM), AChE reactivator pyridine 2-aldoxime methochloride

(2-PAM, 20 mg/kg, IV), and GABAR agonist diazepam (5 mg IV or IM). Other clinically approved AChE reactivating oximes are methoxime, obidoxime, trimedoxime, and HI-6. Some OPs have been reported to cause intermediate syndrome (IMS) in humans. To date, more than a dozen of OPs (bromophos, chlorpyrifos, diazinon, dimethoate, disulfoton, fenthion, malathion, merfos, methamidophos, methyl parathion, monocrotophos, omethoate, parathion, phosmet, and trichlorfon) are known to cause IMS. IMS is usually observed in individuals who have ingested a massive dose of an OP insecticide either accidentally or intentionally in a suicide attempt. IMS is a separate clinical entity from acute toxicity and delayed neuropathy and is characterized by acute respiratory paresis and muscular weakness, primarily in the facial, neck, and proximal limb muscles, occurring 24–96 hours after exposure. IMS appears due to the interaction of OPs with nAChRs at the motor endplate, oxidative stress, and electrophysiological abnormalities at the NMJ.

Some OPs, such as tri-o-cresyl phosphate (TOCP), are known to cause OP-induced delayed polyneuropathy (OPIDPN). Signs and symptoms of this neurodegenerative disorder include tingling of the hands and feet, sensory loss, progressive muscle weakness, flaccidity of the distal skeletal muscles of the extremities, and ataxia. With higher doses of an OP, symptoms can occur as early as 10 days and with lower doses symptoms can take as long as 5 weeks to occur. OPIDPN occurs due to inhibition and aging of neuropathy target esterase (NTE) in peripheral nerves.

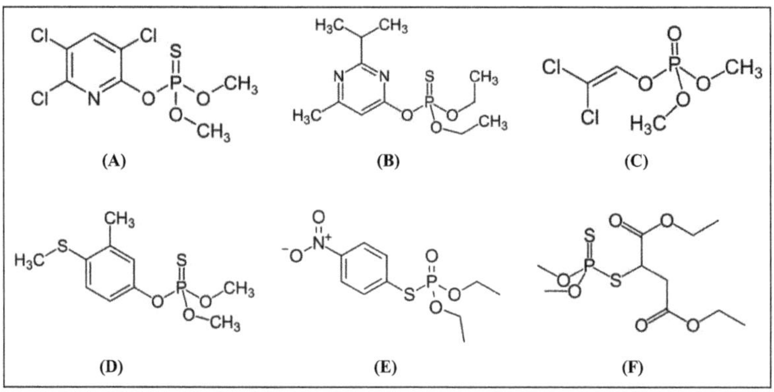

Figure 5.1 Chemical structures of various organophosphate insecticides: **(A)** Chlorpyrifos, **(B)** Diazinon, **(C)** Dichlorvos, **(D)** Fenthion, **(E)** Malathion, and **(F)** Parathion.

5.3.1.2 Carbamate insecticides

Carbamate (CM) insecticides are the esters of methyl or dimethyl carbamic acid. Some commonly used CM insecticides are aldicarb, carbaryl, carbofuran, methomyl, oxamyl, and propoxur, and their chemical structures are shown in Fig. 5.2. CMs produce toxicity by reversible inhibition of AChE. For this reason, CMs are considered safer than OP insecticides. In mild to moderate poisoning cases, recovery can be anticipated within 1–4 hours. Antidotal treatment for CM toxicity rests with atropine sulfate.

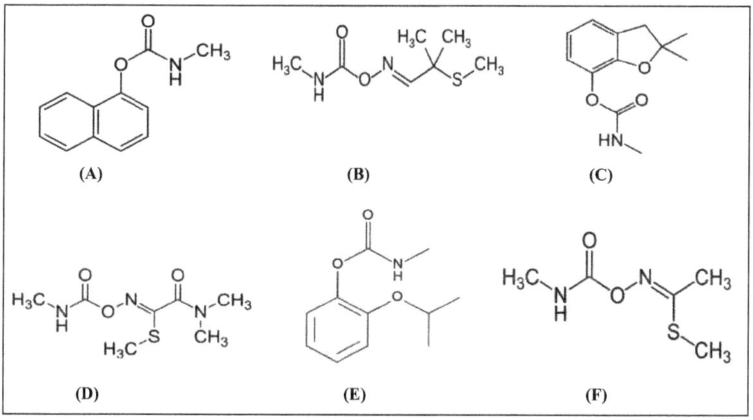

Figure 5.2 Chemical structures of various organophosphate insecticides: **(A)** Aldicarb, **(B)** Carbaryl, **(C)** Carbofuran, **(D)** Methomyl, **(E)** Oxamyl, and **(F)** Propoxur.

5.3.1.3 Organochlorine insecticides

Organochlorine (OC) insecticides include chlorinated ethane derivatives, cyclodienes, and hexachlorocyclohexanes. Some of these chemicals (e.g., DDT) were introduced in the 1940s and were widely used in agricultural and public health programs. This was because of their relatively low acute toxicity and their persistence, which reduced the need for repeated applications. However, their persistence in the biological system and the environment has been since recognized as a liability rather than an asset. Both DDT and methoxychlor are chlorinated ethane derivatives, but methoxychlor is much less toxic and less persistent than DDT. The cyclodiene insecticide endrin is extremely toxic; aldrin and dieldrin are somewhat less toxic; and chlordane, heptachlor, and mirex are even less toxic. Lindane is the gamma isomer of hexachlorocyclohexane (HCH), and it is highly toxic but less cumulative. Consequently, Lindane is more widely used than HCH, especially for hair lice in children. OC insecticides are less water soluble than OP or CM insecticides and more environmentally persistent.

OCs alter the transport of sodium and potassium ions across axonal membranes, resulting in increased negative after-potential and prolonged action potentials. Consequently, repetitive firing occurs after a single stimulus and spontaneous trains of action potentials. Clinical signs and symptoms appear within 24 hours of exposure. The toxic manifestations with chlorinated ethanes and chlorinated benzenes include paresthesia of the tongue, lips, and face, apprehension, tremors, clonic and tonic convulsions, and hyperthermia. With cyclodienes, the signs and symptoms include dizziness, nausea, vomiting, and generalized convulsions. Death may occur within 24–72 hours from respiratory failure due to prolonged paralysis of the respiratory muscles. There is no antidotal treatment.

5.3.1.4 Neonicotinoids

Neonicotinoids are a new class of insecticides with widespread use in crop protection and veterinary medicine. Commonly used neonicotinoids are imidacloprid, acetamiprid, clothianidin, dinotefuran, nitenpyram, thiacloprid, and thiamethoxam. These insecticides produce a neurotoxic insecticidal action by binding to nAChRs of the postsynaptic membrane of nerve cells in the CNS and acting as nAChR agonists.

Neonicotinoids show selective toxicity, that is, lower toxicity in vertebrates compared to insects because they have a lower affinity for binding to nAChRs in mammals than in insects. In mammals with low-dose exposure, the toxic effects include tremors and impaired pupillary function (miosis or mydriasis). At higher doses, death either occurs within 4 hours or recovery takes place within 8–24 hours. There is no specific antidote for neonicotinoid toxicity.

5.3.1.5 Fipronil

Fipronil is a member of a new class of insecticides called phenylpyrazoles, which is used in agriculture and veterinary medicine worldwide. Fipronil and its major metabolites (fipronil sulfone and fipronil desulfinyl) produce neurotoxicity by antagonizing GABAA-regulated chloride channels in the brain. The selective toxicity of fipronil and its metabolites (>500 times in insects over mammals) is due to higher specificity and affinity for receptor binding in insects. Dermal toxicity is low, inhalation toxicity is moderate, and oral toxicity is high. In humans, poisoning is mainly due to accidental ingestion or suicide attempts. The signs and symptoms are hyperactivity including tremors, convulsions, seizures, and death. There is no specific treatment for fipronil toxicity.

5.3.1.6 Amitraz

Amitraz is an amidine insecticide and acaricide used in crop protection and veterinary medicine. Poisoning from amitraz in humans occurs from accidental or intentional ingestion. Amitraz overstimulates α2-adrenoreceptors resulting in general CNS depression, impairment of consciousness, respiratory depression, convulsions, disorientation, miosis, vomiting, bradycardia, hypotension, hypothermia, and hyper/hypoglycemia.

5.3.1.7 Pyrethrins and pyrethroids

Pyrethrins are insecticidal compounds obtained from the flowers of *Chrysanthemum cinerariae*. There are six compounds that comprise natural pyrethrins (pyrethrin I and II, jasmolin I and II, and cinerin I and II). Synthetic pyrethroids have been developed because natural pyrethrins tend to break down quickly when exposed to air, light, and heat. An enzyme inhibitor, piperonyl butoxide, is often used in combination with pyrethrins/pyrethroids as a synergist. The pyrethroids are divided into two types. Type I pyrethroids lack an alpha cyano substituent. Some common examples of this group are allethrin, kadethrin, permethrin, phenothrin, resmethrin, and tetramethrin. Type II pyrethroids contain an alpha cyano substituent. Some examples of this group are cypermethrin, cyphenothrin, deltamethrin, fenpropathrin, fenvalerate, and fluvalinate. Pyrethroids interact directly with voltage-gated sodium channels, thereby interfering with the generation and conduction of nerve impulses. This interference induces pronounced repetitive activity in various parts of the brain. Type I pyrethroids produce tremor (T) syndrome with symptoms of whole-body tremors, incoordination, prostration, tonic-clonic convulsions, and death. Type II pyrethroids elicit choreoathetosis/ salivation (CS) syndrome with symptoms of hyperactive behavior, profuse salivation, tremors, motor incoordination, and death. Both the central nervous system (CNS) and peripheral nervous system (PNS) are affected by pyrethroids. There is no antidotal treatment for pyrethrin/pyrethroid toxicity.

5.3.1.8 Rotenone

Rotenone (nicouline) is one of the oldest naturally occurring compounds present in several plants (*Derris elliptica, Derris involuta, Lonchocarpus nicou, Lonchocarpus urucu, Lonchocarpus utilis, Mundulea sericea,* and *Tephrosia virginiana*). In general, rotenone is an organic pesticide used in home gardens for insect control, for lice and ticks on pets, and for fish eradication as part of water-body management. Rotenone has selective toxicity, as it is highly toxic to insects and fish in comparison to mammalian species. Rotenone converts to highly toxic metabolites in large quantities in insects and fish, while it converts to less toxic metabolites in mammals. Also, rotenone is rapidly absorbed in fish compared to mammalian species. In mammals and fish, it inhibits the oxidation of NADH to NAD+, thereby blocking the oxidation of NAD+ and substrates such as glutamate, alpha-ketoglutarate, and pyruvate. Rotenone causes inhibition of the mitochondrial respiratory chain complex I and cell death by apoptosis due to the excess generation of free radicals. With acute exposure, rotenone produces pharyngitis, nausea, vomiting, gastric pain, clonic convulsions, muscle tremors, lethargy, incontinence, and cardiorespiratory failure. In experimental animals, it is a neurotoxicant and produces signs similar to Parkinson's disease.

5.3.1.9 Nicotine

Nicotine is obtained from the plants *Nicotiana tabacum* and *Nicotiana rustica*. Black Leaf 40® (nicotine sulfate 40%) has been used as an insecticide for more than half a century. Nicotine is known to affect the nervous, cardiovascular, respiratory, endocrine systems, and the gastrointestinal tract. Nicotine is highly toxic to mammals; its pharmacological effects mimic certain effects of ACh and are called "nicotinic effects." Overstimulation of nAChRs at the neuromuscular junction (NMJ) results in tremors, fasciculations, and paralysis of muscles. These effects are blocked by d-tubocurarine. Higher dose nicotine exposure often results in death. Nicotine also affects neuroeffector junctions of the parasympathetic system including the iris of the eye, bladder, heart, tear glands, and salivary glands; these effects are blocked by atropine sulfate.

5.3.2 Herbicides

Herbicides, commonly referred to as "weed killers," are chemicals used to destroy unwanted plants. There are several types of herbicides; some retard the growth of weeds by inhibiting photosynthesis, respiration, cell division, or synthesis of protein or lipids, while others act as growth stimulants, disturbing normal growth. Most human poisoning cases are from accidental exposure to herbicides or suicidal intent. Chlorophenoxy compounds are exemplified by 2,4-dichlorophenoxyacetic acid (2,4-D) and 2,4,5-trichlorophenoxyacetic acid (2,4,5-T). These compounds act as plant growth hormones and have a relatively low toxicity in mammals. By having an additional chlorine, 2,4,5-T is more toxic than 2,4-D. However, chloracne, the main toxic effect of 2,4,5-T in humans, seems attributable to the contaminant 2,3,7,8-tetrachlorodibenzo-p-dioxin (TCDD or dioxin). There is no specific treatment for chlorophenoxy herbicide poisoning.

Bipyridyl herbicides (paraquat and diquat) are widely used in agriculture and horticulture. Life-threatening pulmonary toxicity occurs from occupational or accidental exposure or suicidal attempts. Lungs accumulate paraquat about 10 times more than other tissues, mostly in type I and type II epithelial cells. The mechanism of action of bipyridyl involves cyclic reduction–oxidation reactions that deplete NADPH and generate reactive oxygen species (ROS), which cause cell death by polymerizing unsaturated lipids in cell membranes. Clinical signs include vomiting, abdominal pain, diarrhea, and hypotension, as well as peripheral neuromuscular effects that may persist in surviving victims. Dinitrophenol herbicides include 2,4-dinitrophenol, 2,4-dinitro-o-cresol, and dinoseb. These compounds produce toxicity by blocking oxidative phosphorylation in tissues, which results in stimulation of the metabolism. These compounds also stimulate tissue respiration and impair ATP synthesis. As a result, there is an increase in oxygen demand and an increase in respiration and heart rate, leading to hyperthermia, acidosis, anoxia, and death.

Herbicides of the urea and thiourea group (diuron, flumeturon, isoproturon, linuron, and others) are commonly used in agriculture. These herbicides have low toxicity and do not pose serious risks to humans; however, they may cause eye and skin irritation. Common herbicides of the triazine group include atrazine, simazine, prometon, and propazine. These compounds are of low mammalian toxicity and pose no human risk. The clinical signs of toxicity may include depression, anorexia, weight loss, gastrointestinal upset, tenseness, and weakness.

Carbamate herbicides include chloropropham, carbetamide, desmedipham, and asulam, and thiocarbamate herbicides include dimepiperate, molinate, eradicane, diallate, thiobencarb, and so forth. These herbicides have low mammalian toxicity and pose little human risk.

5.3.3 Fungicides

Fungicides are agents that are used to prevent or eradicate fungal infections from plants or seeds. In agriculture, they are used to protect tubers, fruits, and vegetables during storage and are applied directly to ornamental plants, trees, field crops, cereals, and turf grasses. The inorganic and metallic fungicides include formulations from sulfur, copper, mercury, and potassium. Acute poisoning from sulfur fungicides is very rare and, therefore, does not pose a toxicological threat to mammals. Vineyard sprayers usually develop liver disease after 3–15 years of using fungicidal copper sulfate in a Bordeaux mixture. Mercury compounds such as methyl and ethyl mercury are very effective fungicides, which are widely used to preserve seed grains. However, several tragic accidents involving numerous deaths and permanent neurological damage occurred with their use. In humans, methylmercury disrupts CNS developmental processes in the unborn and infant. These facts have deterred further use of these compounds. Carbamic acid derivative fungicides include ferbam, thiram, ziram, maneb, mancozeb, metiram, nabam, zineb, and so forth. These compounds have been widely used in agriculture and have relatively low mammalian acute toxicity. However, there is concern about their carcinogenic potential.

Chloroalkylthiodicarboximide fungicides have very low acute and chronic toxicity and include captan, captafol, and folpet. Captan, however, was found to reduce the body weight of pups and folpet-induced keratosis of the gastrointestinal tract in rats. Halogenated substituted monocyclic aromatics, such as pentachlorophenol (PCP), have been widely used as wood preservatives. PCP increases metabolic rate through uncoupling of oxidative phosphorylation. It has a low LD50, but its technical grade is more toxic, indicating greater toxicity of its contaminants. Pentachloronitrobenzene (PCNB) has been used as a fungicide to treat soil. It is somewhat less acutely toxic than PCP but may be carcinogenic. Hexachlorobenzene has been used as a seed treatment but has caused mass poisonings as well. Other commonly used fungicides in agriculture include benomyl and thiobendazole, which have very low toxicity. Benomyl is known to induce teratogenic effects, such as tetra amelia, eye defects, miscarriages, and infants with congenital anomalies.

5.3.4 Rodenticides

Rodenticides are chemical agents that kill rodents and are divided into two categories: anticoagulants and non-anticoagulants. Anticoagulant rodenticides are of two types: (1) coumarin type, such as dicoumarol, warfarin, coumachlor, coumafuryl, coumatetralyl, brodifacoum, and bromadiolone; and (2) indanedione type, such as pindone, diphacinone, and chlorophacinone. The mode of action of these rodenticides is twofold: (1) inhibition of prothrombin formation because of interferences with the action of vitamin K in the synthesis of clotting factors (II, VII, IX, and X), and (2) capillary damage resulting in internal bleeding. All mammalian and bird species are susceptible to these compounds, although toxicity varies widely from species to species. In humans, adults are exposed to these compounds because of malicious or suicidal intent, while infants are exposed via accidental ingestion of poisoned baits, and the outcome is usually fatal. Generally, in acute cases, hemorrhage occurs in the cerebral vasculature, pericardial sac, and thorax, followed by convulsions and ataxia. In subacute cases, in addition to hematemesis, bleeding occurs through all body orifices. Antidotal treatment includes vitamin K1 (5 mg/kg body weight, IM, SC, or IV) and whole blood transfusion (20 ml/kg body weight, IV).

Of the many thioureas, alpha naphthyl thiourea (ANTU) has the best rodenticidal effect. ANTU stimulates the sympathetic nervous system and causes a significant increase in permeability of the lung capillaries, and consequently, excessive pulmonary edema and pleural effusion. ANTU is extremely toxic to rats, but only moderately toxic to humans. Sodium fluoroacetate (Compound 1080) and fluoroacetamide (Compound 1081) are extremely toxic and their use has been restricted to licensed personnel. In the body, these compounds get converted to toxic metabolites (fluorocitrate), which exert a toxic effect via blockage of the citric acid cycle by inhibiting the activity of aconitase. These compounds cause damage to the kidneys, heart, and CNS, resulting in death.

Zinc phosphide is another rodenticide that has been used for a long time because it is very effective and inexpensive. The toxicity of zinc phosphide is mainly due to phosphine (PH3), which is formed following a hydrolytic reaction in the stomach. The onset of toxicity is rapid, within a few minutes to a few hours. Clinical signs include vomiting, anorexia, rapid and deep respiration, followed by terminal hypoxia. Bromethalin is a unique, highly potent, single-feeding rodenticide developed in 1985. It was developed as a rodenticide to overcome rodent resistance to certain anticoagulants. Poisoning in humans occurs because of suicidal intent. Bromethalin produces toxicity by uncoupling oxidative phosphorylation in the mitochondria of the CNS. Decreased ATP production inhibits the activity of Na^+/K^+-ATPase and leads to an accumulation of fluid within myelin lamellae and the cerebroventricular system. The increased intracranial pressure damages axons, inhibiting neural transmission, and leads to paralysis, convulsions, and death within 1–3 days.

Other rodenticides include plant alkaloids, such as strychnine and red squill. Strychnine is obtained from the seeds of *Strychnos nux vomica* and *S. ignatii*. Strychnine selectively antagonizes the inhibitory action of glycine in Renshaw cells of the spinal cord and medulla. Clinical signs include rigidity of skeletal muscles, tonic seizures, and death due to paralysis of the respiratory muscles. Red squill is a red powder obtained from the dried bulbs of the sea onion (*Urgenia maritima*) having the active principle scilliroside. Clinical signs include convulsions, hyperesthesia, and incoordination; death occurs due to cardiac toxicity.

5.3.5 Fumigants

As the name implies, this group of pesticides includes several gases, liquids that readily vaporize, and solids that release gases by chemical reactions. Methyl bromide, chloropicrin, and methyl iodide are broad-spectrum fumigants. Other fumigants include dazomet, formaldehyde, hydrogen cyanide, iodoform, methyl isocyanate, phosphine, and sulfuryl fluoride. Fumigants are used in buildings, soil, and grain and are produced to eradicate pests. Fumigation is a hazardous operation and is conducted by certified operators. These chemicals are hazardous to most forms of life, including humans, and inhalation is the primary route of exposure.

5.4 Bioaccumulation and Biomagnification

Bioaccumulation and biomagnification properties do not necessarily represent adverse biological effects. They are generally associated with substances that are lipophilic and resistant to breakdown. The OC pesticides are generally more persistent in the environment and tend to be stored in the fat depot. However, bioaccumulation is more marked with some chemicals than others. For example, DDT is stored in body fat for a much longer time than methoxychlor. The half-lives of these insecticides in rats are 6–12 months and 12 weeks, respectively. The persistence of pesticides in the environment may create ecological problems. DDT and related chemicals in the environment enhance the metabolism of estrogens in birds. In the egg-laying and nestling cycle of certain birds, this disturbance of hormones adversely affects the reproduction and survival of the young. There is evidence that the estrogenic actions of DDT are responsible for the development of breast cancer and disorders of female reproductive functions. Biomagnification is the result of bioaccumulation in the organism alone or in conjunction with persistence in the environment. For example, DDT is lipophilic and occurs in the fat portion of body fluids, including milk. While a mother's daily intake of DDT may be 0.5 μg/kg, her breastfed baby's daily DDT intake may be 11.2 μg/kg. This magnification results from the fact that DDT is stored in the human body at 10–20 times the chronic daily intake level and that the baby consumes, essentially, only milk. Furthermore, the caloric intake per kilogram body weight is higher in babies compared with that of adults. The significance of a greater intake of DDT by babies is not clear because of the relatively short duration of this increased level of intake and the sensitivity of young versus adults. The biomagnification is even more marked in carnivorous animals.

DDT and methylmercury can accumulate through a series of plankton, small fish, large fish, and birds, resulting in a magnification of the concentration amounting to several hundred-fold. The decision to suspend the use of DDT was partly based on its adverse ecological impact, as noted earlier.

Chapter 6

Heavy Metal Toxicity

"Understanding the weight of heavy metals"

6.1 Introduction to Heavy Metals

Heavy metals are metallic elements with a comparatively high density in comparison to water. With the assumption that heaviness and toxicity are linked, heavy metals include metalloids such as arsenic, which can cause toxicity at low levels of exposure. In recent years, there has been an increase in ecological and worldwide public health concerns over these metals and environmental pollution. Human exposure has also increased considerably due to an exponential expansion in their usage in a variety of industrial, agricultural, residential, and technical applications. Geogenic, industrial, agricultural, pharmaceutical, and home effluents and atmospheric sources have all been identified as sources of heavy metals in the environment. Mining, foundries, smelters, and other metal-based industrial processes all contribute significantly to environmental pollution.

Heavy metals have been shown to influence cellular organelles and components such as the cell membrane, mitochondria, lysosomes, endoplasmic reticulum, nuclei, and several enzymes involved in metabolism, detoxification, and damage repair in biological systems. Metal ions have been discovered to interact

Environmental Toxicology: Concepts and Principles
Mohd Salim Reshi, Darakhshan Javaid, Syed Sanober Qadri, and Neelu Sinha

ISBN 978-981-5129-68-7 (Hardcover), 978-1-003-63707-3 (eBook)
www.jennystanford.com

with cell components such as DNA and nuclear proteins, resulting in DNA damage and conformational changes that may lead to cell cycle regulation, cancer, or apoptosis. Several studies have shown that reactive oxygen species (ROS) generation and oxidative stress play an important role in the toxicity and carcinogenicity of metals such as arsenic, cadmium, chromium, lead, and mercury. Because of their extreme toxicity, these five elements are considered priority metals with significant public health implications. They are all systemic toxins that have been shown to cause numerous organ harm at low levels of exposure. These metals are classified as "known" or "probable" human carcinogens by the United States Environmental Protection Agency (USEPA) and the International Agency for Research on Cancer (IARC) based on epidemiological and experimental studies, showing an association between exposure and cancer incidence in humans and animals.

6.2 Metals of Major Toxicological Concern

Although heavy metals are naturally occurring elements and found throughout the earth's crust, most environmental contamination and human exposure result from anthropogenic activities such as mining and smelting operations, industrial production and use, and domestic and agricultural use of metals and metal-containing compounds. Environmental contamination can also occur through metal corrosion, atmospheric deposition, soil erosion of metal ions, leaching of heavy metals, sediment re-suspension, and metal evaporation from water resources to soil and groundwater. Natural phenomena such as weathering and volcanic eruptions have also been reported to significantly contribute to heavy metal pollution. Industrial sources include metal processing in refineries, coal burning in power plants, petroleum combustion, nuclear power stations, high-tension lines, plastics, textiles, microelectronics, and wood and paper processing plants.

It has been reported that metals such as cobalt (Co), copper (Cu), chromium (Cr), iron (Fe), magnesium (Mg), manganese (Mn), molybdenum (Mo), nickel (Ni), selenium (Se), and zinc (Zn) are essential nutrients that are required for various biochemical and physiological functions. An inadequate supply of these micronutrients results in a variety of deficiency diseases or syndromes as shown in (Fig. 6.1).

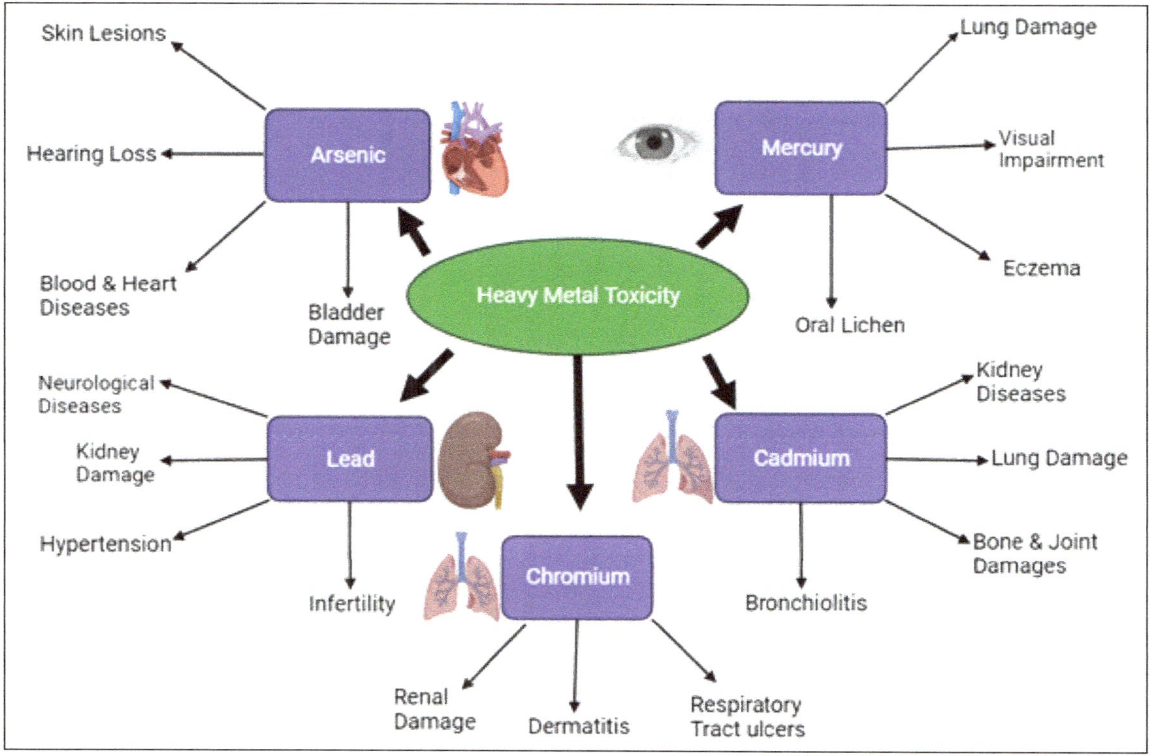

Figure 6.1 Metals having high toxicity and their effects on humans.

6.2.1 Arsenic

6.2.1.1 Environmental occurrence, industrial production, and use

Arsenic is a ubiquitous element that is detected at low concentrations in virtually all environmental matrices. The major inorganic forms of arsenic include the trivalent arsenite and the pentavalent arsenate. The organic forms are the methylated metabolites – monomethylarsonic acid (MMA), dimethylarsinic acid (DMA), and trimethylarsine oxide. Environmental pollution by arsenic occurs because of natural phenomena such as volcanic eruptions, soil erosion, and anthropogenic activities. Several arsenic-containing compounds are produced industrially and have been used to manufacture products with agricultural applications such as insecticides, herbicides, fungicides, algicides, sheep dips, wood preservatives, and dye-stuffs. They have also been used in veterinary medicine for the eradication of tapeworms in sheep and cattle. Arsenic compounds have also been used in the medical field for at least a century in the treatment of syphilis, yaws, amoebic dysentery, and trypanosomiasis. Arsenic-based drugs are still used in treating certain tropical diseases such as African sleeping sickness and amoebic dysentery, and in veterinary medicine to treat parasitic diseases, including filariasis in dogs and blackhead in turkeys and chickens. Recently, arsenic trioxide has been approved by the Food and Drug Administration as an anticancer agent in the treatment of acute promyelocytic leukemia. Its therapeutic action has been attributed to the induction of programmed cell death (apoptosis) in leukemia cells.

6.2.1.2 Potential for human exposure

It is estimated that several million people are exposed to arsenic chronically throughout the world, especially in countries like Bangladesh, India, Chile, Uruguay, Mexico, and Taiwan, where the groundwater is contaminated with high concentrations of arsenic. Exposure to arsenic occurs via the oral route (ingestion), inhalation, dermal contact, and the parenteral route to some extent. Arsenic concentrations in air range from 1 to 3 ng/m^3 in remote locations (away from human releases), and from 20 to 100 ng/m^3 in cities. Its water concentration is usually less than 10 μg/L, although higher levels can occur near natural mineral deposits or mining sites. Its concentration in various foods ranges from 20 to 140 mg/kg. Natural levels of arsenic in soil usually range from 1 to 40 mg/kg, but pesticide application or waste disposal can produce much higher values.

Diet, for most individuals, is the largest source of exposure, with an average intake of about 50 μg per day. Intake from air, water, and soil is usually much smaller, but exposure from these media may become significant in areas of arsenic contamination. Workers who produce or use arsenic compounds in occupations such as vineyards, ceramics, glass-making, smelting, refining of metallic ores, pesticide manufacturing and application, wood preservation, and semiconductor manufacturing can be exposed to substantially higher levels of arsenic. Arsenic has also been identified at 781 sites of the 1,300 hazardous waste sites that have been proposed by the USEPA for inclusion on the national priority list. Human exposure at these sites may occur by a variety of pathways, including inhalation of dust in the air, ingestion of contaminated water or soil, or through the food chain.

Contamination with high levels of arsenic is of concern because arsenic can cause several human health effects. Several epidemiological studies have reported a strong association between arsenic exposure and increased risks of both carcinogenic and systemic health effects. Interest in the toxicity of arsenic has been heightened by recent reports of large populations in India, Bangladesh, Thailand, Mongolia, Taiwan, China, Mexico, Argentina, Chile, Finland, and Hungary that have been exposed to high concentrations of arsenic in their drinking water and are displaying various clinicopathological conditions including cardiovascular and peripheral vascular disease, developmental anomalies, neurologic and neurobehavioral disorders, diabetes, hearing loss, portal fibrosis, hematologic disorders (anemia, leukopenia, and eosinophilia), and carcinoma. Arsenic exposure affects virtually all organ systems including cardiovascular, dermatologic, nervous, hepatobiliary, renal, gastrointestinal, and respiratory systems. Research has also pointed to significantly higher standardized mortality rates for cancers of the bladder, kidney, skin, and liver in many areas of arsenic pollution. The severity of adverse health effects is related to the chemical form of arsenic and is also time- and dose-dependent. Although the evidence of the carcinogenicity of arsenic in humans seems strong, the mechanism by which it produces tumors in humans is not completely understood.

6.2.2 Cadmium

6.2.2.1 Environmental occurrence, industrial production, and use

Cadmium is a heavy metal of considerable environmental and occupational concern. It is widely distributed in the earth's crust at an average concentration of about 0.1 mg/kg. The highest level of cadmium compounds in the environment is accumulated in sedimentary rocks, and marine phosphates contain about 15 mg of cadmium per kilogram.

Cadmium is frequently used in various industrial activities. The major industrial applications of cadmium include the production of alloys, pigments, and batteries. Although the use of cadmium in batteries has shown considerable growth in recent years, its commercial use has declined in developed countries in response to environmental concerns. In the United States, for example, the daily cadmium intake is about 0.4 μg/kg/day, less than half of the USEPA's oral reference dose. This decline has been linked to the introduction of stringent effluent limits from plating works and, more recently, to general restrictions on cadmium consumption in certain countries.

6.2.2.2 Potential for human exposure

The main routes of exposure to cadmium are via inhalation or cigarette smoke, and ingestion of food. Skin absorption is rare. Human exposure to cadmium is possible through several sources including employment in primary metal industries, eating contaminated food, smoking cigarettes, and working in cadmium-contaminated workplaces, with smoking being a major contributor. Other sources of cadmium include emissions from industrial activities, including mining, smelting, and manufacturing of batteries, pigments, stabilizers, and alloys. Cadmium is also present in trace amounts in certain foods such as leafy vegetables, potatoes, grains and seeds, liver and kidney, and crustaceans and mollusks.

In addition, foodstuffs, that are rich in cadmium, such as liver, mushrooms, shellfish, mussels, cocoa powder, and dried seaweed, can greatly increase the cadmium concentration in human bodies. An important distribution route is the circulatory system; whereas, blood vessels are mainstream organs of cadmium toxicity. Chronic inhalation exposure to cadmium particulates is generally associated with changes in pulmonary function and chest radiographs that are consistent with emphysema. Workplace exposure to airborne cadmium particulates has been associated with a decrease in olfactory function. Several epidemiologic studies have documented an association of chronic low-level cadmium exposure with a decrease in bone mineral density and osteoporosis.

Exposure to cadmium is commonly determined by measuring cadmium levels in blood or urine. Blood cadmium reflects recent cadmium exposure (from smoking, for example). Cadmium in urine (usually adjusted for dilution by calculating the cadmium/creatinine ratio) indicates accumulation or kidney burden of cadmium. It is estimated that about 2.3% of the U.S. population has elevated levels of urine cadmium (>2 µg/g creatinine), a marker of chronic exposure and body burden. Blood and urine cadmium levels are typically higher in cigarette smokers, intermediate in former smokers, and lower in non-smokers. Because of the continuing use of cadmium in industrial applications, the environmental contamination and human exposure to cadmium have dramatically increased during the past century.

6.2.3 Chromium

6.2.3.1 Environmental occurrence, industrial production, and use

Chromium (Cr) is a naturally occurring element present in the earth's crust, with oxidation states (or valence states) ranging from chromium (II) to chromium (VI). Chromium compounds are stable in the trivalent [Cr (III)] form and occur in nature in this state in ores, such as ferrochromite. The hexavalent [Cr (VI)] form is the second-most stable state. Elemental chromium [Cr (0)] does not occur naturally. Chromium enters various environmental matrices (air, water, and soil) from a wide variety of natural and anthropogenic sources with the largest release coming from industrial establishments. Industries with the largest contribution to chromium release include metal processing, tannery facilities, chromate production, stainless steel welding, and ferrochrome and chrome pigment production. The increase in the environmental concentrations of chromium has been linked to air and wastewater release of chromium, mainly from metallurgical, refractory, and chemical industries. Chromium released into the environment from anthropogenic activity occurs mainly in the hexavalent form [Cr (VI)]. Hexavalent chromium [Cr (VI)] is a toxic industrial pollutant that is classified as a human carcinogen by several regulatory and non-regulatory agencies. The health hazard associated with exposure to chromium depends on its oxidation state, ranging from the low toxicity of the metal form to the high toxicity of the hexavalent form. All Cr (VI)-containing compounds were once thought to be man-made, with only Cr (III) naturally ubiquitous in air, water, soil, and biological materials. Recently, however, naturally occurring Cr (VI) has been found in ground and surface waters at values exceeding the World Health Organization limit for drinking water of 50 μg of Cr (VI) per liter. Chromium is widely used in numerous industrial processes and as a result, is a contaminant of many environmental systems. Commercially chromium compounds are used in industrial welding, chrome plating, dyes and pigments, leather tanning, and wood preservation. Chromium is also used as anticorrosive in cooking systems and boilers.

6.2.3.2 Potential for human exposure

It is estimated that more than 300,000 workers are exposed annually to chromium and chromium-containing compounds in the workplace. In humans and animals, [Cr (III)] is an essential nutrient that plays a role in glucose, fat, and protein metabolism by potentiating the action of insulin. However, occupational exposure has been a major concern because of the high risk of Cr-induced diseases in industrial workers occupationally exposed to Cr (VI). Also, the general human population and some wildlife may be at risk. It is estimated that 33 tons of total Cr are released annually into the environment. The U.S. Occupational Safety and Health Administration (OSHA) recently set a "safe" level of 5 $\mu g/m^3$, for an 8-hour time-weighted average, even though this revised level may still pose a carcinogenic risk. For the general human population, atmospheric levels range from 1 to 100 ng/cm^3 but can exceed this range in areas that are close to Cr manufacturing.

Non-occupational exposure occurs via ingestion of chromium-containing food and water whereas occupational exposure occurs via inhalation. Chromium concentrations range between 1 mg/kg and 3000 mg/kg in soil, 5 μg/L and 800 μg/L in seawater, and 26 μg/L and 5.2 mg/L in rivers and lakes. Chromium content in foods varies greatly and depends on the processing and preparation. In general, most fresh foods typically contain chromium levels ranging from <10 to 1,300 μg/kg. Present-day workers in chromium-related industries can be exposed to chromium concentrations two orders of magnitude higher than the general population. Even though the principal route of human exposure to chromium is through inhalation, and the lung is the primary target organ, significant human exposure to chromium has also been reported to take place through the skin. For example, the widespread incidence of dermatitis noticed among construction workers is attributed to their exposure to chromium present in cement. Occupational and environmental exposure to Cr (VI)-containing compounds is known to cause multiorgan toxicity such as renal damage, allergy and asthma, and cancer of the respiratory tract in humans.

Breathing high levels of chromium (VI) can cause irritation to the lining of the nose, and nose ulcers. The main health problems seen in animals following ingestion of chromium (VI) compounds are irritation and ulcers in the stomach and small intestine, anemia, sperm damage, and male reproductive system damage. Chromium (III) compounds are much less toxic and do not appear to cause these problems. Some individuals are extremely sensitive to chromium (VI) or chromium (III), and allergic reactions consisting of severe redness and swelling of the skin have been noted. An increase in stomach tumors was observed in humans and animals exposed to chromium (VI) in drinking water. Accidental or intentional ingestion of extremely high doses of chromium (VI) compounds by humans has resulted in severe respiratory, cardiovascular, gastrointestinal, hematological, hepatic, renal, and neurological effects. Although the evidence of carcinogenicity of chromium in humans and terrestrial mammals seems strong, the mechanism by which it causes cancer is not completely understood.

6.2.4 Lead

6.2.4.1 Environmental occurrence, industrial production, and use

Lead is a naturally occurring bluish-grey metal in small amounts in the earth's crust. Although lead occurs naturally in the environment, anthropogenic activities such as burning fossil fuels, mining, and manufacturing contribute to the release of high concentrations. Lead has many different industrial, agricultural, and domestic applications. It is currently used in the production of lead-acid batteries, ammunition, metal products (solder and pipes), and devices to shield X-rays. An estimated 1.52 million metric tons of lead were used for various industrial applications in the United States in 2004. Of that amount, lead-acid battery production accounted for about 83%, and the remaining usage covered a range of products such as ammunition (3.5%), oxides for paint, glass, pigments, and chemicals (2.6%), and sheet lead (1.7%).

In recent years, the industrial use of lead has been significantly reduced from paints and ceramic products to sealing and pipe solders. Despite this progress, it has been reported that among 16.4 million homes with more than one child younger than 6 years per household in the United States, 25% of homes still had significant amounts of lead-contaminated deteriorated paint, dust, or adjacent bare soil. Lead in dust and soil often re-contaminates cleaned houses and contributes to elevated blood lead concentrations in children who play on bare, contaminated soil. Today, the largest source of lead poisoning in children comes from dust and chips from deteriorating lead paint on interior surfaces. Children who live in homes with deteriorating lead paint can achieve blood lead concentrations of 20 μg/dL or greater.

6.2.4.2 Potential for human exposure

Exposure to lead occurs mainly via inhalation of lead-contaminated dust particles or aerosols, and ingestion of lead-contaminated food, water, and paints. Adults absorb 35–50% of lead through drinking water and the absorption rate for children may be greater than 50%. Lead absorption is influenced by factors such as age and physiological status. In the human body, the greatest percentage of lead is taken into the kidney, followed by the liver and the other soft tissues such as the heart and brain, however, the lead in the skeleton represents the major body fraction. The nervous system is the most vulnerable target of lead poisoning. Headache, poor attention span, irritability, loss of memory, and dullness are the early symptoms of the effects of lead exposure on the central nervous system.

Since the late 1970s, lead exposure has decreased significantly because of multiple efforts including the elimination of lead in gasoline, and the reduction of lead levels in residential paints, food, drinking cans, and plumbing systems. Several federal programs implemented by state and local health governments have not only focused on banning lead in gasoline, paint, and soldered cans but have also supported screening programs for lead poisoning in children and lead abatement in housing. Despite the progress in these programs, human exposure to lead remains a serious health problem. Lead is the most systemic toxicant that affects several organs in the body including the kidneys, liver, central nervous system, hematopoietic system, endocrine system, and reproductive system.

Lead exposure usually results from lead in deteriorating household paints, lead in workplaces, lead in crystals and ceramic containers that leach into water and food, lead used in hobbies, and lead used in some traditional medicines and cosmetics. Several studies conducted by the National Health and Nutrition Examination Surveys (NHANES) have measured blood lead levels in U.S. populations and have assessed the magnitude of lead exposure by age, gender, race, income, and degree of urbanization. Although the results of these surveys have demonstrated a general decline in blood lead levels since the 1970s, they have also shown that large populations of children continue to have elevated blood lead levels (> 10 µg/dL). Hence, lead poisoning remains one of the most common pediatric health problems in the United States today. Exposure to lead is of special concern among women, particularly during pregnancy. Lead absorbed by the pregnant mother is readily transferred to the developing fetus. Human evidence corroborates animal findings, linking prenatal exposure to lead with reduced birth weight and preterm delivery, and with neuro-developmental abnormalities in offspring.

6.2.5 Mercury

6.2.5.1 Environmental occurrence, industrial production, and use

Mercury is a heavy metal belonging to the transition element series of the periodic table. It is unique in that it exists or is found in nature in three forms (elemental, inorganic, and organic), with each having its profile of toxicity. At room temperature, elemental mercury exists as a liquid, which has a high vapor pressure and is released into the environment as mercury vapor. Mercury also exists as a cation with oxidation states of +1 (mercurous) or +2 (mercuric). Methylmercury is the most frequently encountered compound of the organic form found in the environment. This is formed due to the methylation of inorganic (mercuric) forms of mercury by microorganisms found in soil and water.

Mercury is a widespread environmental toxicant and pollutant that induces severe alterations in the body tissues and causes a wide range of adverse health effects. Both humans and animals are exposed to various chemical forms of mercury in the environment. These include elemental mercury vapor (Hg^0), inorganic mercurous (Hg^{+1}), mercuric (Hg^{+2}), and organic mercury compounds. Because mercury is ubiquitous in the environment, humans, plants, and animals are all unable to avoid exposure to some form of mercury.

Mercury is utilized in the electrical industry (switches, thermostats, and batteries), dentistry (dental amalgams), and numerous industrial processes including the production of caustic soda, in nuclear reactors, as antifungal agents for wood processing, as a solvent for reactive and precious metal, and as a preservative of pharmaceutical products. The industrial demand for mercury peaked in 1964 and sharply declined between 1980 and 1994 because of federal bans on mercury additives in paints and pesticides, and the reduction of its use in batteries.

Humans are exposed to all forms of mercury through accidents, environmental pollution, food contamination, dental care, preventive medical practices, industrial and agricultural operations, and occupational operations. The major sources of chronic, low-level mercury exposure are dental amalgams and fish consumption. Mercury enters water as a natural process of off-gassing from the earth's crust and through industrial pollution. Algae and bacteria methylate the mercury entering the waterways. Methyl mercury then makes its way through the food chain into fish, shellfish, and eventually into humans.

The two most highly absorbed species are elemental mercury (Hg^0) and methyl mercury (MeHg). Dental amalgams contain over 50% elemental mercury. The elemental vapor is highly lipophilic and is effectively absorbed through the lungs and tissues lining the mouth. After Hg^0 enters the blood, it rapidly passes through cell membranes, which include both the blood-brain barrier and the placental barrier. Once it enters the cell, Hg^0 is oxidized and becomes highly reactive to Hg^{2+}. Methyl mercury derived from eating fish is readily absorbed in the gastrointestinal tract. Because of its lipid solubility, it can easily cross both the placental and blood-brain barriers. Once mercury is absorbed, it has a very low excretion rate. A major proportion of what is absorbed accumulates in the kidneys, neurological tissue, and the liver. All forms of mercury are toxic and their effects include gastrointestinal toxicity, neurotoxicity, and nephrotoxicity.

Chapter 7

Toxicity Tests and Antidotal Therapy

"From poison to remedy: the science of recovery"

7.1 Introduction

Toxicology is a Greek word that means "the science of poison." In other words, it is a branch of pharmacology that deals with various aspects of poisons and poisoning. Any substance that, even in small doses, produces adverse effects on the metabolism of an organism and consequently may cause death is Toxin.

Du Bois and Geiling (1959) defined toxicology as a branch of medical science that deals with the nature, properties, effects, and detection of poison.

7.2 Toxicity Testing

Toxicity testing is paramount in the screening of newly developed drugs before they can be used on humans. The essence of toxicity testing is not just to check how safe a test substance is, but to characterize the possible toxic effects it can produce. The guiding principles of toxicity testing are to check the effect of the test

Environmental Toxicology: Concepts and Principles
Mohd Salim Reshi, Darakhshan Javaid, Syed Sanober Qadri, and Neelu Sinha

ISBN 978-981-5129-68-7 (Hardcover), 978-1-003-63707-3 (eBook)
www.jennystanford.com

substances on laboratory animals and their direct toxic effect on humans and secondly, the exposure of laboratory animals to high doses to evaluate its possible hazard to humans that are exposed to much lower doses.

Toxicity testing employed a wide range of tests in different species of animals with long-term administration of drugs, regular monitoring of physiological and biochemical abnormalities, and detailed post-mortem examination at the end of the trial to detect gross or histological abnormalities.

7.3 Types of Toxicity Testing

Toxicity testing is of the following types:

- Acute toxicity
- Sub-acute toxicity
- Chronic toxicity

7.3.1 Acute Toxicity

Acute toxicity tests can provide preliminary information on the toxic nature of a material for which no other toxicology information is available. This is a short-term assessment and evaluation of potential hazard test substance or consequences of a single dose of a test substance.

Acute toxicity testing may be used in risk assessments of chemicals for humans and non-target environmental organisms. Acute toxicity study is better described as LD_{50}, which is defined as the dose that kills 50% of animals. LD_{50} values have been used to compare relative acute hazards of industrial chemicals, especially when no other toxicology data are available for the chemicals. Rats are used as the animal model in acute toxicity tests, but other species also may be used. Often only one sex is studied in an acute toxicity test; generally, the female is assumed to be more sensitive to the acute toxic effects of chemicals than the male. Before deciding on the dose of a test compound that will be used in studying its acute toxicity, the compound's chemical and physical characteristics (including molecular weight, partition coefficient, and the toxicity of related chemicals) should be considered; otherwise, oral toxicity including lethality caused by relatively large doses of a chemical may have no biological relevance to the chemical's effects at lower doses.

The formula for calculating the LD_{50} is expressed as:

$$LD_{50} = \frac{[M_0 + M_1]}{2}$$

M_0 = Highest dose of test substance without mortality.

M_1 = Lowest dose of test substance that gave mortality.

7.3.1.1 Types of acute toxicity

- **Limit Tests**: To determine the acute toxicity of a new food additive that is not expected to be particularly toxic, 5 gm (or ml) of the compound per kilogram body weight of the test animal should be administered orally by gavage to several (perhaps 5) animals that have been fasted (overnight for rats, 4 hours for mice). Test animals should be observed closely for up to 14 days; symptoms of toxicity and recovery should be noted. Gross and histopathological examination of the test animals at the end of the study may help identify toxic effects on target organs. If no animals die because of this dose, there is no need to test higher dosages. The acute toxicity of the compound can then be expressed as being greater than 5 gm (or ml)/kg body weight of the test animal. This method is called the "limit test." In general, 5 gm or 5 ml of the test substance per kilogram body weight is the practical upper limit for test material that can be administered in one oral gavage dose to a rodent. If there are deaths following administration of an acute dose of 5 gm/kg body weight, then a lower dose should be administered to several (perhaps 5) animals, and the results evaluated as discussed above. For compounds expected to be acutely toxic at 5 gm/kg body weight, it would be wise to select a lower initial "limit" dose.
- **Dose-Probing Tests**: Dose-probing acute toxicity protocols may have value when the petitioner has no preliminary information about the test substance that would help select appropriate doses for toxicity studies. In a dose-probing acute toxicity test, one animal per each of 3 widely spaced dosages should be used and a sufficient observation period should follow the administration of the doses. Subsequent toxicity studies may be based on the results of the dose probing study.

- **Up-and-Down Tests**: This procedure involves dosing animals one at a time. First, one animal at one dose; then another animal, one or two days later at a higher dose (if the first animal survives) or a lower dose (if the first animal dies). This process continues until the approximate LD_{50} has been determined. One disadvantage to this test is the length of the study. Each animal should be observed for at least 7 days after dosing so that delayed deaths can be recorded. However, this method usually requires only 6–8 test animals compared to 40–50 test animals that may be used in the "classical" LD_{50} test.
- **Pyramiding Tests**: This involves a minimum number of animals. Two animals are given successively increasing doses of the test substance on alternate days until an acutely toxic dose or some practical upward limit is reached. This test does not yield a lethality curve and is used to assess acute toxicity in non-rodents. This test, although more like a short-term, repeated dose toxicity study than a true acute toxicity study, can provide useful preliminary information on the toxic nature of new materials for which no other toxicology information is available.

7.3.1.2 Importance of toxicity studies

- To establish a dose-response curve.
- To ensure the safety of new chemicals for use as pesticides, drugs, or food additives before they are registered for general use in industry or doctors' clinics.
- To establish the mode of action or mechanism for a toxic effect that may have been seen in other studies.
- To produce epidemiological studies, to explain observations in the population, for instance, the long investigation into the association of smoking with lung.
- To validate new methods of testing or investigation, particularly those conducted in vitro rather than in animals.
- To deal with cases of accidental ingestion of a large amount of the material (e.g., for poison control information).
- To determine possible target organs that should be scrutinized and/or special tests that should be conducted in repeated-dose toxicity tests.

- To select doses for short-term and sub-chronic toxicity tests when no other toxicology information is available.

7.3.2 Sub-acute Studies

It is also known as short-term toxicity. Sub-acute toxicity tests are intended to evaluate the toxicity of the chemical after repeated administration and to help establish doses for longer-term sub-chronic studies. Most sub-acute studies utilize 3–4 different dosages of chemicals, administered by mixing them in the feed. For rodent studies, 10 animals of each sex are usually used at each dose; whereas, for dogs, 3 dosages and 3–4 animals per sex are used. The chemical is typically administered for 14 days, after which the animals are killed and complete clinical chemistry and histopathology analyses are performed.

Objectives of sub-acute toxicity tests:

- To assess the nature of the toxic effects.
- Human beings are often exposed to chemicals at much lower levels but over a longer period of time.

7.3.2.1 Sub-chronic toxicity testing

Sub-chronic toxicity assessment requires information obtained from both the acute and sub-acute toxicity test levels. This will help determine the test doses to be used at this level. At least, 3 dose levels are recommended for this test. The test doses can be determined using the following formula:

$$\text{Dose} = M_0 / [1+ k]$$

where, M_0 = Highest dose of test substance without mortality after the sub-acute test, and

k = positive real numbers starting from 0 with an increment of 1 (i.e., $k = 0$, $k = 1$, and $k = 2$ for three respective dose levels).

Ten animals (consisting of equal numbers of two different strains) are required for each group. A control group of 10 (consisting of equal numbers of two different strains) is also required. The test doses are administered to the animals in the respective groups for a period of 30 days while the control group animals are administered with the vehicle. Observation of all animals is done on a daily basis throughout the period of study. Signs of toxicity and mortality are recorded. The mortality of two animals from the same strain in a particular group indicates toxicity at that dose. At any time, mortality is observed, the vital organs (kidney, lung, liver, heart, spleen, small intestine, stomach, duodenum, and adrenal gland) of such animals are isolated immediately for histological examination. At the end of the period of administration, four animals (two from each strain) from each group (including the control group) are sacrificed 6 hours after the last administration. The vital organs (kidney, lung, liver, heart, spleen, small intestine, stomach, duodenum, and adrenal gland) are isolated for histological examination. The histology of the test group's organs is compared with the control. Where there is mortality in a test group within the period of assessment, the dead animal(s) is counted as among the sacrificed. The histology is compared with the control. Signs of organ tissue damage if observed are recorded.

7.3.3 Chronic Toxicity Testing

It is also known as long-term but definite toxicity. Chronic toxicity assessment can only be carried out for test doses that pass the sub-chronic test. Chronic toxicity tests are defined as tests that characterize adverse effects following repeated administration of a test substance over a significant portion of the lifespan of the test species. Establishing the duration of a chronic study is based on the anticipated human or environmental species exposure. For example, when pharmaceuticals are tested under FDA guidance, the length of exposure in animals is established based on the portion of the human lifespan over which the test substance is expected to be used.

It begins 24 hours after administration of the last sub-chronic doses. Six animals each (consisting of three animals per strain) should be alive from both control and passing dose groups at the end of the sub-chronic procedure (including histological assessment). Administration of the test doses to the animals in the respective groups is continued for a period of 62 days (making a total of 92 days of administration), while the control group animals still receive the vehicle. All animals in the respective groups are carefully observed daily throughout the period of study. Weight-taking across the groups is also done at a 2-week interval. Mortality and other signs of toxicity are recorded. At any time, mortality is observed, the vital organs (kidney, lung, liver, heart, spleen, small intestine, stomach, duodenum, and adrenal gland) of such animal(s) should be isolated immediately for histological examination. The date (with respect to the test) of such mortality is recorded. The mortality of two animals from the same strain in a particular group shows toxicity at that dose. At the end of the period of administration, all the animals from the test group control are sacrificed 6 hours after the last administration. The vital organs (kidney, lung, liver, heart, spleen, small intestine, stomach and duodenum, and adrenal gland) are isolated for histological examination. The histology of the test group's organs is compared with the control. Where there is mortality in a test group within the period of assessment, the animal is counted among the sacrificed, and the histology is compared with the control. Signs of organ tissue damage are recorded.

Objectives of chronic toxicity test:

- To study the adverse effects of chemicals on the structure and function of organs, tissues, and cells after a prolonged period of exposure.
- To study secondary effects of toxicants such as carcinogenicity.
- To evaluate safe levels of toxicants.

7.4 Concluding Remarks and Future Directions

Toxicity assessment is as important as discovering and developing new agents for man's use. Although some attention has been given to the issue of toxicity testing of pharmacological/chemical agents, more efforts still need to be put into it. At present, there are still a number of pharmacological/chemical agents with little or no (or even inaccurate) toxicity information about them.

The toxic effect of the drug can range from negligible to severe as to preclude further development of the compound. The use of animals in toxicity testing is most likely to continue for the foreseeable future because of the benefits they offer in examining a whole functioning organism.

7.5 Antidotal Therapy

The International Programme of Chemical Safety broadly defines an antidote as a therapeutic agent that counteracts the toxic actions of a drug/toxin. Broadly, antidotes have been looked at as agents that "modify the kinetics of the toxic substance or interfere with its effect at receptor sites." This may be because of the prevention of absorption, binding, and neutralizing the poison directly, antagonizing its end-organ effect, or inhibiting the conversion to more toxic metabolites. A chemical's safety is defined by its therapeutic index or ratio (TD_{50}/ED_{50}), which is the ratio of the toxic dose (TD) or lethal dose (LD) to the effective dose (ED). Based on this, an antidote has also been defined as an agent that "increases the mean lethal dose of a toxin."

Hence, they are agents that negate the effect of a poison or toxin. Antidotes mediate its effect either by preventing the absorption of the toxin, binding and neutralizing the poison, antagonizing its end-organ effect, or inhibiting the conversion of the toxin to more toxic metabolites. Examples of mechanisms by which antidotes work include competition at a receptor site, alteration of a metabolic process, engaging a counter-regulatory physiologic process, or hastening the excretion or detoxification of a toxin. Although antidotes are generally beneficial in intoxicated patients, some of them are themselves toxic and should be used only when indicated.

7.5.1 How Does an Antidote Work?

Antidotes, however, have a broader meaning in terms of altering the effect of a toxin. Two main variables impact the harmful effect of a toxin on the body, namely the dose and the duration of exposure to the toxin. These in turn depend on the type of toxin, the dose, the route of administration, lag time to presentation to a hospital, and pharmacokinetics (absorption, distribution, and elimination).

Antidotes work in any one of several ways. Common modes of action are as follows:

1. Inert Complex Formation: Some antidotes interact with the poison to form an inert complex, which is then excreted from the body, e.g., chelating agents for heavy metals, Prussian Blue for thallium, specific antibody fragments for digoxin, etc.

2. **Accelerated Detoxification:** Some antidotes accelerate the detoxification of a poison, e.g., thiosulfate accelerates the conversion of cyanide to non-toxic thiocyanate, acetylcysteine acts as a glutathione substitute, which combines with hepatotoxic paracetamol metabolites and detoxifies them.

3. **Reduced Toxic Conversion:** The best example of this mode of action is provided by ethanol, which inhibits the metabolism of methanol to toxic metabolites by competing for the same enzyme (alcohol dehydrogenase).

4. **Receptor Site Competition:** Some antidotes displace the poison from specific receptor sites, thereby antagonizing the effects completely. The best example is provided by naloxone, which antagonizes the effects of opiates at stereo-specific opioid receptor sites.

5. **Receptor Site Blockade:** This mode of action is best exemplified by atropine, which blocks the effects of anti-cholinesterase agents such as organophosphates at muscarinic receptor sites.

6. **Toxic Effect Bypass:** An example of this type of antidotal action is provided using 100% oxygen in cyanide poisoning.

Table 7.1 A list of genuine antidotes recommended in toxicological practice today

Antidote	Main Indication	Other Applications
N-Acetylcysteine	Paracetamol	Amanitin
Atropine	Cholinergic agents	Amanitin
Ethanol	Methanol, ethylene glycol	Copper, gold, mercury
Dimercaprol	Arsenic	Copper, gold, mercury
Oximes	Organophosphates	Disulfiram, coprin
Hydroxocobalamin	Cyanide	Copper, gold, mercury
Potassium hexacyanoferrate	Thallium	Gold, lead, mercury
Glucose	Insulin	Copper, gold, mercury
Sodium nitrite	Cyanide	Hydrogen sulfide
Calcium salts	Oxalates, fluorides	Calcium antagonists
Penicillamine	Copper	Gold, lead, mercury
Sodium thiosulfate	Cyanide	Bromate, chlorate, iodine

Table 7.1 represents a list of genuine antidotes recommended in toxicological practice today. In addition, there are certain therapeutic agents that are not antidotes as per the accepted definition, but which through their importance and sometimes specific role in the treatment of poisons, border on the concept of "antidotes."

A brief discussion on some of the common specific antidotes is as follows:

1. N-acetylcysteine: N-acetylcysteine (NAC) prevents NAPQI-induced hepatoxicity through several mechanisms. Firstly, it acts as a glutathione (GSH) precursor that increases GSH availability, as well as providing inorganic sulfate that enhances paracetamol metabolism through the non-toxic sulfation pathway. NAC also directly converts NAPQI to relatively non-toxic cysteine and inter-capitate conjugates, besides reducing NAPQI back to paracetamol, which is then eliminated by other non-toxic routes. NAC also offers additional benefits because of its non-specific antioxidant effects and free radical scavenging properties. If administered early in the course of exposure, NAC can prevent significant paracetamol-induced toxicity. Even when given later, it can still ameliorate toxicity. NAC also has a role in limiting toxicity caused by glutathione depletion and free radical formation from poisoning due to carbon tetrachloride, chloroform, pennyroyal oil, and even valproic acid.

2. Atropine: Atropine is a competitive antagonist at both central and peripheral muscarinic receptors, and it is effective in the treatment of exposures to muscarinic agonists and acetylcholinesterase inhibitors such as organophosphate and carbamate pesticides, as well as some chemical warfare agents. Like scopolamine (hyoscine), atropine is a tropane alkaloid found in Datura and some other related plants, and has a tertiary amine structure that allows CNS penetration. This is an advantage that some other anti-muscarinic agents such as glycopyrrolate, ipratropium, and tiotropium do not have, which are unable to cross the blood-brain.

3. Cyanide Antidote:

- *Amyl and sodium nitrites*: Amyl and sodium nitrites are highly effective cyanide antidotes, but they must be administered without delay. Amyl nitrite is a volatile liquid that is available in some Western countries in ampoules that can be broken, and the vapor is inhaled by the patient until intravenous sodium nitrite can be begun the usual dose of sodium nitrite in adults is 300 mg (10 mL of 3% solution), which should be given intravenously at a rate of 2.5–5 mL/min. Half this dose can be repeated if symptoms of cyanide toxicity reappear. If amyl nitrite is available, it should be administered first.

- *Hydroxocobalamin*: Hydroxocobalamin has been used as a cyanide antidote in some European countries, most notably France, for many years, either as a sole agent or in combination with sodium thiosulfate. The usual dose of hydroxocobalamin is 70 mg/kg (to a maximum of 5–10 g initially) administered intravenously over 30 minutes. This dose can be repeated (up to a maximum total dose of 15 g) as necessary. The second and subsequent doses should be administered over a longer period (6–8 hours), except in severe cases.

4. Metal Chelating Agents: Most chelating agents have similar mechanisms of action. Soft metal ions, such as Hg^{2+}, Au^{+}, Cu^{+}, and Ag^{+}, have large ionic radii with a large number of electrons in their outer shell. Therefore, they form the most stable complexes with sulfur donors and are referred to as sulfur seekers. Chelating agents such as BAL (British Anti-Lewisite) form a coordinate bond with the metal by donating a pair of free electrons. Hard metals such as Na^{+}, K^{+}, Mg^{2+}, Ca^{2+}, and Al^{3+} are referred to as oxygen seekers and form complexes with chelators containing a carboxyl (COO^{-}) group, such as calcium disodium edetate (CaNa2EDTA). Borderline metal ions, such as Pb^{2+}, Cd^{2+}, Cu^{2+}, As^{3+}, and Zn^{2+}, prefer nitrogen-donating ligands but will also react with both hard and soft ligands. Antidotes for metal poisoning often contain more than one type of donating group, making them effective for more than one type of metal.

5. Oximes: The most commonly used oxime for organophosphate pesticide (OP) poisoning in India is pralidoxime (2-hydroxyiminomethyl-1-methyl pyridinium chloride, pyridine-2-aldoxime methiodide - P2AM or PAM). The others such as oxime and obidoxime are not available in India. Pralidoxime must invariably be administered together with atropine in the management of OP poisoning.

Organophosphate pesticides are powerful inhibitors of carboxylic esterase enzymes, especially acetylcholinesterase, which is found in red blood cells, nerve cells, and skeletal muscle, as well as plasma cholinesterase or butyrylcholinesterase, which is found in plasma, liver, heart, pancreas, and brain). The former is often referred to as true cholinesterase, and the latter as pseudocholinesterase.

6. Ethanol: Ethanol is used as an antidote in those cases where a particular toxic substance is metabolized by alcohol dehydrogenase, e.g., methanol and ethylene glycol. It acts by competitive inhibition because toxic metabolites of methanol and ethylene glycol are not formed, and the parent compound is excreted unchanged. The affinity of ethanol for alcohol dehydrogenase is 67 times that of ethylene glycol and 15.5 times that of methanol. A relatively smaller quantity of ethanol is required to inhibit the metabolism of ethylene glycol when compared to the amount required to block the metabolism of methanol since the affinity of ethylene glycol for alcohol dehydrogenase is much less than that of methanol. The usual recommendation is to maintain a serum ethanol concentration of 100 mg/dL, or at least a 1:4 molar ratio of ethanol to methanol or ethylene glycol, whichever is greater.

Ethanol can be given orally or intravenously. The usual concentrations recommended for oral and IV administrations are 20–30% and 5–10%, respectively. Intravenous administration is always preferable since it leads to complete absorption, does not produce GI effects, and can be administered to an unconscious patient.

7. Vitamin K: Vitamin K is an essential fat-soluble vitamin, and it is a generic term that includes two distinct natural forms: vitamin K1 (phytonadione, phylloquinonc) and vitamin K2 (menaquinones) comprising a series of compounds with the same 2-methyl-1,4-naphthoquinonering structure as phylloquinone, but with a variable number (1–13) of repeating 5-carbon units on the side chain. Most of the dietary vitamin K is in the form of phytonadione. Phytonadione is antidotal in nature for reversing elevated prothrombin time/ international normalized ratio (INR) induced by toxins or drugs such as warfarin, or long-acting anticoagulant rodenticides (LAARs), such as brodifacoum. It is usually given orally because IV administration of vitamin K1 is associated with anaphylactoid reactions. Subcutaneous administration may be considered if a patient is unable to tolerate oral vitamin K.

The duration of antidotal therapy depends on the type of toxin consumed, the estimated dose that the individual has been exposed to, the route of exposure, clinical features of toxicity, half-life, and pharmacokinetics as well as the risk vs benefit for the use of the antidote.

Chapter 8

Endocrine Disruption: Chemicals and Hormonal Systems

"Disrupted hormones, disrupted lives"

8.1 Historical Perspectives

The initial observations suggesting that hormonal imbalance during pregnancy might lead to abnormal fetal development date back to the 1930s. In 1939, researchers at Northwestern University Medical School documented that when pregnant rats received an additional dose of external estrogen, their offspring experienced structural defects in their sex organs, affecting both females and males. For a considerable period, this phenomenon was considered specific to rodents and not thought to be relevant to humans. Additionally, there was a prevailing belief within the scientific and medical community that the human placenta served as an impervious barrier, protecting the developing fetus from exposure to chemicals encountered by a pregnant woman. The confidence in the concept of a placental barrier was shattered by the tragic incident involving thalidomide.

Environmental Toxicology: Concepts and Principles
Mohd Salim Reshi, Darakhshan Javaid, Syed Sanober Qadri, and Neelu Sinha

ISBN 978-981-5129-68-7 (Hardcover), 978-1-003-63707-3 (eBook)
www.jennystanford.com

8.1.1 The Thalidomide Tragedy and DES Controversy

Thalidomide developed in 1957, gained widespread use in Europe and Australia as a prescription drug for pregnancy-related purposes, serving as a tranquilizer and anti-nausea medication. However, it was subsequently withdrawn from the market due to severe deformities observed in some babies born to women who had taken thalidomide during pregnancy. These deformities included the absence or underdevelopment of limbs. Interestingly, not all babies born to women taking thalidomide experienced such deformities. Instead, the occurrence of deformities was linked to the timing of thalidomide intake, specifically during the organ-forming period between the fifth and eighth week of pregnancy.

Diethylstilbestrol (DES), synthesized as a synthetic estrogen analog in 1943, was widely prescribed to pregnant women for miscarriage prevention in the following decades. However, a 1952 epidemiological study at the University of Chicago found no difference in miscarriage frequency between women who took DES and those who did not. Despite this, physicians continued to prescribe DES through the 1960s. In 1971, two independent case-control epidemiological studies revealed a high frequency of vaginal cancer among girls born to women who took DES, occurring at an unusually young age of 15–22. Although both studies showed impressive statistical significance, some researchers questioned the methodology's validity.

Subsequent studies uncovered a higher incidence of reproductive organ abnormalities, such as T-shaped uteri in women and abnormal testicles, genital tumors, low sperm counts, and abnormal sperm in men, compared to the incidence of vaginal cancer. Additionally, there were indications of a higher-than-usual prevalence of homosexual and bisexual tendencies among women exposed to DES in utero. Structures of thalidomide and diethylstilbestrol are shown in Fig. 8.1.

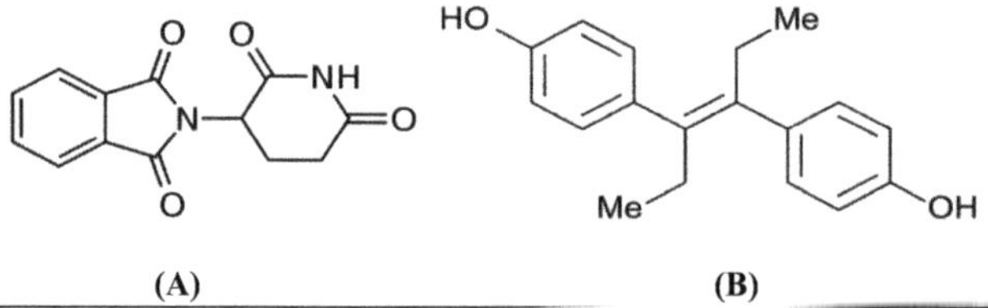

(A) (B)

Figure 8.1 Structures of (A) thalidomide and (B) diethylstilbestrol.

8.2 Hormonal Imbalance

The examples of thalidomide and DES, along with subsequent studies on environmental contaminants, underscore the critical importance of maintaining correct hormonal balance for proper fetal development. Both estrogens and androgens are naturally present in males and females, though in different ratios. Any disruption in the proper balance of sex hormones can lead to abnormal sexual development and potentially skew the ratio of females to males among newborns. Prenatal or early postnatal exposure to an excess of natural hormones or hormone mimickers, as well as antagonists, can disturb the entire endocrine balance of fetuses, resulting in abnormal development.

The significance of maintaining proper hormonal balance during in-utero development was demonstrated by Vom Saal in mice. Female fetuses located in the uterus between two male fetuses exhibited significantly higher concentrations of testosterone in both their blood and amniotic fluid. The adult mice that developed from these fetuses displayed male-like aggressiveness and lacked pheromones that would make them sexually attractive to males. This research highlights the profound impact of hormonal imbalances during fetal development on later behavior and reproductive characteristics.

Our knowledge of what determines whether a fertilized egg becomes a male or a female is very recent. Essentially the process appears to be very simple. The eggs produced by the mother carry an X chromosome. The sperm produced by a father may contain either an X or a Y chromosome. If a sperm carrying an X chromosome combines with an egg the resulting embryo will be a female. If a Y chromosome carrying sperm fertilizes an egg and the embryo will be male, the developing embryo is not committed for some time whether it will develop as a male or a female. The final issue depends on hormonal cues received during embryonic development. It is then likely that hormonal imbalance may result in altered sex ratio, or abnormal sexual development leading, in extreme cases, to hermaphrodites. Although most studies were done with compounds interacting with estrogen receptors, androgen and thyroid functions could be affected. The name endocrine disrupters had been coined for compounds interfering with hormonal balance.

8.3 Properties of Endocrine Disrupters

A considerable number of structurally and functionally unrelated compounds have been identified as endocrine disruptors. These compounds not only lack structural resemblance to the hormones they mimic or interfere with, but they also often lack structural similarity among themselves. In contrast to carcinogens, where a structure–activity relationship is identifiable, no such relationship exists among compounds with hormonal activities.

Colborn et al. identified 44 chemicals widely dispersed in the environment with endocrine-disrupting activities. These included herbicides, fungicides, insecticides, nematicides, and industrial products or by-products such as certain heavy metals, PCBs, dioxins, plasticizers (alkylphenols and bisphenol-A, Fig. 8.2), and more. Many of these compounds are resistant to degradation, fat-soluble, and possess a high vapor pressure, making them easily transportable through air circulation.

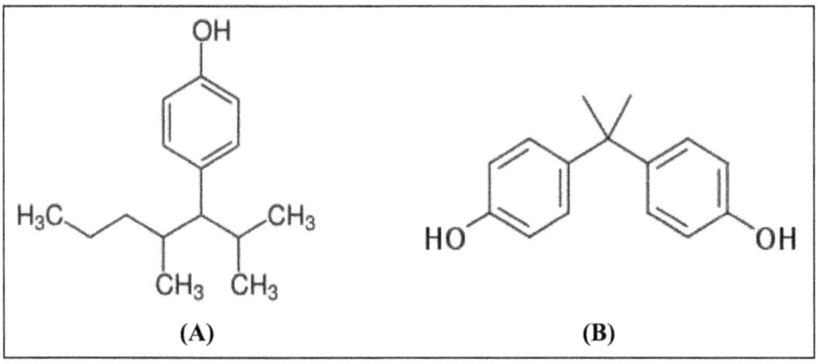

Figure 8.2 Structures of (A) nonylphenol, a plasticizer added to polystyrene plastics, and (B) bisphenol-A, a plasticizer added to polycarbonate plastics.

Endocrine disrupters can exert their effects through direct or indirect mechanisms. Direct-acting disrupters interact with hormonal receptors either by mimicking the natural hormone or by inhibiting its action. On the other hand, indirect-acting disrupters interfere with the synthesis of sterol, the precursor of sex hormones. These disrupters can operate at levels comparable to natural hormones, often at parts per trillion. Some may require activation by the xenobiotic metabolizing system.

While the plasma levels of natural hormones are finely regulated by binding any excess to plasma proteins, rendering the hormone temporarily inactive, the plasma proteins are unable to bind to hormonal mimics. This lack of binding increases the effective dose of the mimic, even if it is less potent than the natural hormone or occurs at a lower concentration. Many environmental endocrine disrupters exhibit an abnormal dose–response curve, taking the shape of an inverted U. This means that the disrupters may have an optimal range of effectiveness, and their impact might decrease at higher or lower concentrations.

What makes the endocrine disrupters especially insidious is that the damage they cause is irreversible and may appear only several years after exposure. Moreover, the damage is not limited to the reproductive system. In both sexes, the internal and external genitalia, brain, skeleton, thyroid, liver, kidney, and immune systems are potential targets.

8.4 Environmental and Health Impact of Endocrine Disrupters

Following the end of World War II, a significant influx of various chemicals entered the environment. Some, like pesticides and chemical fertilizers, were intentionally applied, while others were industrial products or by-products that escaped accidentally or were disposed of in unprotected pits. Increasing reliance on chemicals was viewed as a sign of progress.

It was not until 1962 that Rachel Carson's book, *Silent Spring*, brought attention to the dangers of chemical contamination in the environment, with a primary focus on the concerns about cancer and human health. Since then, many persistent, fat-soluble compounds such as chlorinated pesticides (DDT, aldrin, chlordane, toxaphene) or industrial products (PCBs) have been banned in the United States and most industrialized countries. However, their legacy persists as remnants from previous use or are transported by air currents from countries where they are still in use.

Research on endocrine disrupters spanned decades, with knowledge limited to the scientific community. It was not until the publication of the book *Our Stolen Future* by Colborn et al. in 1996 that awareness of the potential environmental and health impacts of endocrine disrupters reached governmental authorities, if not the public.

Cases of wildlife damage caused by endocrine disrupters are widespread globally. While a definitive causal link is not established in all cases, a consistent pattern emerges: polluted water corresponds to sickened or abnormally developed animals, including mammals, fish, birds, or reptiles.

8.4.1 Fish and Fish-Eating Birds

In 1981, Moccia et al. noted thyroid pathology in the Great Lakes coho and chinook salmon, which appeared to have environmental etiology. Also, cases of feminization were reported to occur in male fish. It has been observed in some rivers in the United Kingdom that male fish, which congregated in the vicinity of effluents from wastewater treatment plants developed vitellogenin, a protein that normally occurs only in female fish and is needed for the development of egg yolk. The authors of this report tested several chemicals known to be estrogenic in mammals and have shown that they were also estrogenic in fish. Many of these estrogenic chemicals are known to occur in effluents from wastewater purification plants. A more recent study conducted in the United Kingdom demonstrated that the compounds responsible for the feminization of male fish were natural and synthetic estrogens (estradiol and ethinyl estradiol, respectively). These compounds, which are the main active ingredient of the contraceptive pill, were found to be present in a very low concentration of effluents from wastewater purification plants. An extensive review provides historical data to support the hypothesis that organochlorine chemicals introduced into the Great Lakes after World War II are the cause of reproductive failures among bald eagles, which was presented by Colborn. An association between pollution of the Great Lakes with organochlorine compounds, primarily PCB, and immunosuppression in pre-fledgling Caspian terns and herring gulls was also reported. A study by scientists for the International Joint Commission reported embryo mortality, edema, and deformities syndrome in colonial fish-eating birds from the Great Lakes. Indirect evidence suggested a causal relation between the observed syndromes and pollution of Lake Ontario with TCDD. The evidence of this causal relation was further reinforced by the fact that the improvement of reproduction of Lake Ontario herring gulls coincided with the decline in organochlorine compounds, particularly TCDD and PCB. Similarly, Donaldson et al. suggested that the recovery of the bald eagle population in the Canadian Great Lakes is related, among other factors, to a reduction in organochlorine levels in the waters of the Great Lakes. The detrimental effect of organochlorine compounds on fish-eating birds could be further evidenced by an experiment in which gull eggs were injected with DDT at concentrations comparable to those found in contaminated seabird eggs. Male birds that hatched from these eggs showed signs of feminization; and development of ovarian tissue and oviducts.

8.4.2 Mollusks

Masculinization of females has been documented in various mollusk species in marinas and harbors in the United Kingdom, along the Connecticut coast in the United States, and even offshore in the middle of the North Sea within shipping lanes. The root cause of this abnormal sexual development has been identified as the presence of tributyl tin (TBT) in the paint used on boats and ships. TBT is an antifouling compound designed to prevent the accumulation of barnacles. This observation highlights the unintended consequences of certain chemicals used in marine environments, leading to disruptive effects on the reproductive biology of marine organisms.

8.4.3 Marine Mammals

In 1988, scientists from the Université de Montréal conducted necropsies on stranded beluga whales in highly polluted areas of the St. Lawrence River. Two of the carcasses showed severe multisystemic lesions, with one exhibiting severe necrotizing dermatitis associated with Herpes-like particles. Additionally, four other whales displayed five varieties of tumors. The presence of high concentrations of benz(a)pyrene-DNA adducts correlated with the elevated incidence of tumors. These animals also exhibited high concentrations of organochlorine compounds in their tissues. Other reports also documented tumors in the St. Lawrence beluga whale population, emphasizing the health risks associated with long-term exposure to pollutants in their environment and suggesting that they might serve as a model to predict health problems in highly exposed human populations over time.

Reproductive failure was observed among common seals feeding on fish from polluted waters off the coast of the Netherlands. Between 1950 and 1975, the seal population declined significantly from over 3000 to less than 500 animals. The authors attributed this reproductive failure to varying levels of polychlorinated biphenyls (PCBs) in the tissues of seals from the western (the Netherlands) and northern parts of the Wadden Sea, with significantly different PCB levels. The die-off of North Sea seals in 1988 was attributed to suppressed immunity. While a distemper-like virus was identified as the primary cause, healthy animals were resistant to it. However, compromised immunity, likely due to exposure to endocrine disrupters, made the animals susceptible to the virus. This underscores the intricate relationship between environmental pollutants, immune system function, and disease susceptibility in marine mammal populations.

8.4.4 Reptiles

In 1980, an extensive spill of the pesticide dicofol occurred in Florida at the Tower Chemical Company located in the vicinity of Lake Apopka. The site of the spill was placed on the EPA Superfund list. Examination of Lake Apopka alligators revealed that 6-month-old, female alligators had plasma estrogen concentrations almost two times greater than alligators from uncontaminated lakes. They also exhibited abnormal ovarian morphology. The male alligators had depressed testosterone concentration, poorly organized testes, and abnormally small penises. Abnormal sexual development in both males and females was probably responsible for the reproductive failure of the species.

8.4.5 Terrestrial Mammals

Many remaining members of the declining population of the Florida Panther have been found to exhibit a range of abnormalities, including low sperm count, abnormal sperm, thyroid dysfunction, immunosuppression, and congenital heart defects. While these defects were initially attributed to the lack of genetic diversity in the population, current evidence suggests that environmental pollutants, such as DDE, PCB, and mercury, may be significant factors contributing to the decline of the species. This underscores the potential impact of environmental contaminants on the health and reproductive success of wildlife populations, even in the absence of other genetic factors.

8.4.6 Humans

The effect of endocrine disrupters in humans is not as well documented as in wildlife. However, in a few cases, the causal relation has been clearly shown; whereas, in others, there was a high likelihood that such a relation existed. The most significant study in this area was done by Jacobson et al. These researchers compared 242 babies born to mothers who consumed fish from Lake Michigan, contaminated with PCBs, with 71 control infants whose mothers did not eat fish. The experimental group exhibited subtle behavioral impairments, such as motor immaturity, a greater startle response, and more abnormally weak reflexes, as compared to the controls. There was also an inverse relationship between the mother's fish consumption and the baby's birth weight and head circumference. Several IQ and achievement tests were administered to the same children when they were 11 years of age. The exposed children showed lower IQ than the controls and were at least two years behind in reading comprehension. Canadian health officials noted that many children in Inuit people villages in arctic Quebec suffered from chronic ear infections and that there were abnormalities in their immune systems. There might be a correlation between these symptoms and the traditional diet of people inhabiting the Arctic region, which consists largely of fish and marine mammals. A few studies revealed heavy dietary uptake of organochlorine compounds and a considerable contamination of the mother's milk with PCBs, DDE, and dieldrin. Contamination of the arctic marine food chain with organochlorine compounds was due to the atmospheric transfer of these compounds from the industrialized regions further to the south. Studies from France, Denmark, and the United States indicated a steady decrease in sperm count and motility among men in the United States and industrialized countries of Europe. The Parisian study (32) revealed a 2.1% per year decrease in sperm concentration and a 0.6% decrease in motility among Parisian men between 1973 and 1992. The American study compared the sperm concentration decreases by regions. They reported that the decline in sperm count was seen in the United States and Europe, but not in non-Western countries. The Danish researcher hypothesized that the increasing incidence of reproductive abnormalities in men might be related to exposure to endocrine disrupters in utero and proposed a mechanism of action.

Chapter 9

Medical Toxicity and Pathology

"From toxic exposure to pathological insights"

9.1 Introduction

Understanding how toxins interact with biological systems and the resulting pathological changes is not only crucial for diagnosis and treatment but also essential for preventing potential harm. Biological agents often encountered in our environment, have the potential to disrupt normal physiological processes, leading to a spectrum of clinical manifestations. Through this exploration, we aim to provide healthcare professionals, researchers, and students with a robust foundation for understanding and navigating the complexities of medical toxicity.

9.2 Definition

In medical terminology, toxicity refers to the degree to which a substance (such as a drug, chemical, or biological agent) can harm an organism. This harm may be caused by the substance itself, its metabolites (substances produced as it is processed by the body), or a combination of both.

Environmental Toxicology: Concepts and Principles
Mohd Salim Reshi, Darakhshan Javaid, Syed Sanober Qadri, and Neelu Sinha

ISBN 978-981-5129-68-7 (Hardcover), 978-1-003-63707-3 (eBook)
www.jennystanford.com

Toxicity can vary depending on factors such as the dose, duration of exposure, route of exposure (e.g., ingestion, inhalation, skin contact), and individual susceptibility. It can manifest as a range of adverse effects, from mild symptoms to severe illness, organ damage, or even death.

Understanding the toxicity of substances is crucial in various fields of medicine, including pharmacology (the study of drugs), occupational health, environmental health, and toxicology (the study of poisons and their effects). It helps healthcare professionals make informed decisions about the use of substances and take appropriate measures to prevent or mitigate harm.

Toxic agents encompass a wide range of substances that have the potential to harm living organisms. These agents can be categorized into several main groups, including:

1. **Chemical Toxic Agents:** These include various industrial chemicals, household products, and agricultural pesticides. Examples of chemical toxic agents include lead, mercury, arsenic, and solvents like benzene and toluene. Toxic heavy metals like lead, mercury, cadmium, and arsenic can cause significant harm when exposure occurs, often through contamination of water, food, or the environment. Toxic gases, such as chlorine, ammonia, and carbon monoxide, can be released from industrial processes, accidents, or deliberate acts.

2. **Pharmaceutical Drugs:** Certain drugs can be toxic if taken in excessive doses or if a patient is sensitive to them. For example, an overdose of medications like acetaminophen or non-steroidal anti-inflammatory drugs (NSAIDs) can lead to liver or kidney damage. Some cancer drugs, while necessary for treatment, are highly toxic and can cause side effects.

3. **Biological Agents:** Bacteria, viruses, fungi, and parasites can produce toxins that cause illnesses. Examples include bacterial toxins like botulinum toxin (produced by *Clostridium botulinum*) or hepatitis viruses. Toxins produced by animals like snakes, spiders, scorpions, and marine creatures can be highly toxic and lead to envenomation when they bite or sting.

4. **Environmental Pollutants:** Pollutants like particulate matter, volatile organic compounds (VOCs), and nitrogen dioxide can have harmful effects on the respiratory and cardiovascular systems.

Contaminants in water sources, such as heavy metals, pesticides, and industrial chemicals, can be ingested, leading to health issues.

Soil contamination with substances like lead, cadmium, or organic pollutants can affect the quality of crops and groundwater.

5. Radiation: X-rays, gamma rays, and radioactive materials can cause damage to cells and DNA, potentially leading to cancer and other health problems. Ultraviolet (UV) radiation from the sun, microwave radiation, and radiofrequency radiation from electronic devices can also be harmful in excess.

6. Food and Beverage Contaminants: Bacteria like *Salmonella* and *E. coli*, as well as toxins produced by these microorganisms, can cause food poisoning. Pesticide residues, food additives, and naturally occurring toxins in food and beverages can be toxic when consumed in large quantities.

9.3 Absorption, Distribution, Metabolism, and Excretion (ADME)

In human toxicology and pathology, the absorption, distribution, metabolism, and excretion (ADME) of toxicants play a crucial role in understanding how these substances affect the body (Fig. 9.1).

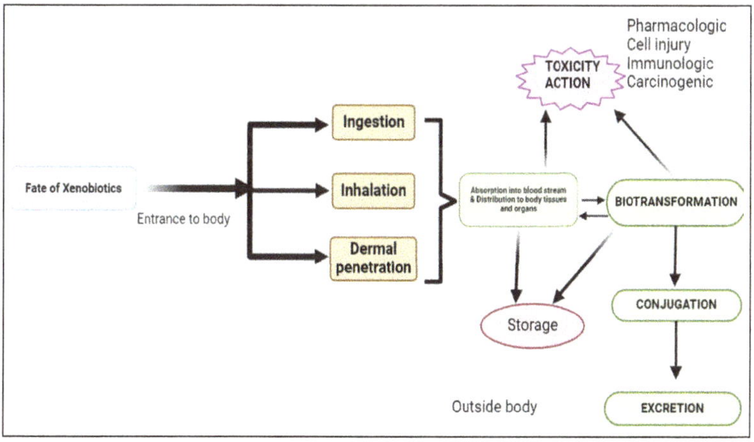

Figure 9.1 Fate of xenobiotics.

1. Absorption: This is the initial step where a toxicant enters the body. Absorption can occur through various routes, such as ingestion, inhalation, dermal contact, or injection. The extent and rate of absorption depend on factors like the chemical properties of the toxicant, the route of exposure, and individual factors like the person's age, gender, and health status. For example, toxicants inhaled as gases or vapors can be rapidly absorbed through the lungs and enter the bloodstream, while substances ingested must pass through the digestive system.

2. Distribution: After absorption, toxicants are distributed throughout the body via the bloodstream. The extent of distribution depends on the toxicant's solubility in blood and various tissues. Some toxicants may preferentially accumulate in specific tissues or organs, while others may be more evenly distributed. Factors influencing distribution include blood flow, the binding of the toxicant to proteins in the blood, and the toxicant's ability to cross cellular membranes.

3. Metabolism (Biotransformation): Many toxicants undergo biotransformation, primarily in the liver, where they are metabolized into more polar compounds that can be more easily excreted from the body. This process can make the toxicants less or more toxic, depending on the specific reactions involved. Enzymes in the liver, such as the cytochrome P450 system, play a critical role in this phase. Some toxicants become reactive intermediates during metabolism, which can bind to cellular components and lead to tissue damage. The outcome of metabolism varies from one toxicant to another.

4. Excretion: The final step is the elimination of toxicants from the body. Excretion can occur through various routes, including urine, feces, sweat, and exhalation. The kidneys play a central role in excreting water-soluble toxicants, primarily through urine, while less polar or fat-soluble compounds are often eliminated through bile in the feces. The rate of excretion depends on the toxicant's chemical properties and its interactions with various excretory organs.

Understanding the ADME of toxicants is essential in toxicology and pathology to assess the potential risks, toxic effects, and therapeutic interventions for individuals exposed to harmful substances. It can also aid in the development of strategies to minimize exposure and prevent toxic outcomes in both acute and chronic poisoning cases.

9.4 Organ-Specific Toxicity

9.4.1 Hepatotoxicity

Hepatotoxicity refers to the occurrence of liver damage caused by exposure to chemicals, drugs, or other toxic substances. The liver, a key organ in metabolizing and detoxifying substances, is particularly vulnerable to injury from these agents. Hepatotoxicity can manifest in various forms, including inflammation, necrosis, cholestasis, fibrosis, and even liver failure. This condition is a significant concern in the fields of medicine and toxicology due to the liver's essential role in maintaining overall health and metabolic function. Several factors contribute to the liver's susceptibility to xenobiotic toxicity.

Firstly, when chemicals are ingested orally, the liver, due to its anatomical proximity to the gastrointestinal tract, serves as the initial site for processing these xenobiotics.

Secondly, the portal blood, which perfuses the intestines, carries not only chemicals and nutrients but also byproducts from the breakdown of intestinal microorganisms, such as inflammogenic lipopolysaccharide components (endotoxin). Increased endotoxin delivery during xenobiotic intoxication may intensify the hepatotoxicity of ingested chemicals, particularly affecting heavy alcohol drinkers.

Thirdly, besides entering through the portal circulation, chemicals can also reach the liver through arterial blood, mixing with venous blood in the hepatic sinusoids. For instance, tobacco constituents inhaled through the lungs efficiently travel to the liver via the arterial route.

Fourthly, the liver's extensive metabolic capabilities paradoxically increase its vulnerability to chemical toxicity. In a miniature chemical factory, hepatic enzymes like CYPs (cytochrome P450 enzymes) can unintentionally generate harmful metabolites, leading to "bioactivation-dependent" hepatotoxicity.

1. Basic Liver Structure: The gross anatomy of the human liver reveals four unevenly sized lobes enveloped by a thin connective tissue capsule known as Glisson's capsule. The liver enjoys a dual blood supply, with approximately 80% of the blood entering through the portal vein, which collects poorly oxygenated blood from the intestines—this is the route through which orally ingested xenobiotics reach the liver. The remaining blood supply consists of well-oxygenated arterial blood conveyed by the hepatic artery. These blood vessels access the liver at the porta hepatis, where efferent bile ducts and lymphatic vessels exit the organ. The venous drainage of the liver takes place through the inferior vena cava.

2. Hepatocytes: The liver is composed of various cell types, with the predominant and crucial cell being the hepatocyte, which constitutes approximately four-fifths of the organ's volume (Fig. 9.2). Hepatocytes serve as the primary workforce of the liver, executing numerous vital functions. Among their key roles, hepatocytes extract nutrients from the blood, produce bile to aid food digestion and fat absorption, expel metabolic waste into the bile, synthesize proteins for both hepatic and non-hepatic purposes, and store energy-rich substances like fat and glycogen.

In the context of toxicology, hepatocytes play a central role in detoxifying lipophilic xenobiotics. They express significant levels of crucial xenobiotic-metabolizing enzymes, including CYPs, UGTs, SULTs, NATs, and GSTs. These enzymes enable hepatocytes to influence the toxicological fate of various xenobiotics. Regardless of a candidate drug's promising pharmacological properties in bioassays, any inclination to transform into harmful metabolites during metabolic processing by hepatocytes can significantly impact its viability as a pharmaceutical for human use.

Hepatocytes are pivotal in the toxicological field, influencing the destiny of many xenobiotics. The development of methods to culture hepatocytes from laboratory animals and human liver samples has been a landmark achievement in toxicology. Primary hepatocyte cultures, with appropriate media and supplements, can replicate a broad spectrum of xenobiotic biotransformations, mirroring the metabolic alterations that occur in the liver in vivo. Consequently, hepatocyte cultures serve as essential tools in research and screening during studies of xenobiotic toxicity.

3. Sinusoids: In the intact liver, hepatocytes are arranged in layers or laminae, forming a distinctive three-dimensional lattice. This parenchymal tissue mass consists of a network of small blood vessels known as sinusoids, which both nourish the hepatocytes and remove metabolic wastes (refer to Fig. 9.2). The blood carried by sinusoids within the liver is a blend of venous and arterial blood, the composition of which depends on the contractility of the supplying blood vessels.

Sinusoids are lined by sinusoidal endothelial cells (SECs), accounting for nearly half of the sinusoidal volume. These SECs create an unusually 'leaky' barrier that functions as a sieve. This barrier restricts the interaction of erythrocytes and circulating immune cells with hepatocytes while facilitating the rapid uptake of small blood-borne substances, including lipophilic xenobiotics, by hepatocytes. This distinctive architecture ensures the efficient exchange of substances between blood and hepatocytes while maintaining a level of control over the types of interactions that occur within the liver.

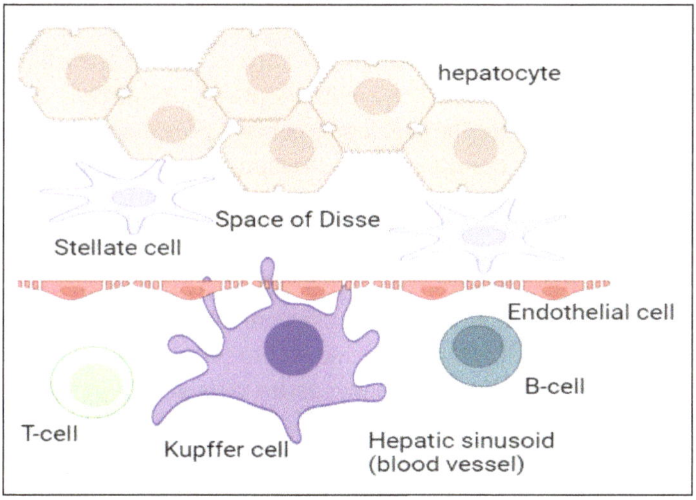

Figure 9.2 Cross-section through a hepatic sinusoid (blood vessel) shows a layer of hepatocytes running parallel to SECs (sinusoidal endothelial cells) within the vascular wall. Other liver cell types that participate in hepatotoxic responses include the Kupffer cells and the stellate cells.

4. Lobules and Metabolic Zonation: The cellular organization within the liver lattice is traditionally conceptualized through the liver lobule, a polygonal structure bordered at its periphery by terminal branches of the hepatic artery and portal vein. The lobular center is defined by the central venule. Within the hepatic lobule, hepatocytes align linearly along sinusoidal tracts, resulting in spatial variations in oxygenation and metabolic function. This arrangement creates three main microenvironments for hepatocytes: centrilobular, midlobular, and periportal zones.

These distinct zones exhibit significant heterogeneity in xenobiotic metabolism. Centrilobular hepatocytes typically express higher levels of CYPs and UGTs, while SULTs are primarily expressed in periportal liver cells. Morphological differences in the smooth endoplasmic reticulum between centrilobular and periportal hepatocytes correspond to variations in CYP expression. Additionally, differences in oxygen tension across the liver lobule influence the toxicity of certain chemicals.

This heterogeneity ensures that hepatotoxicants often damage cells in specific zones. For example, allyl alcohol and cocaine predominantly injure periportal hepatocytes, while toxicants requiring CYP-catalyzed bioactivation, such as carbon tetrachloride (CCl_4) and paracetamol (acetaminophen), typically induce centrilobular damage. The liver lobule concept provides a framework for understanding the differential susceptibility of hepatocytes to various toxic insults based on their spatial distribution within the liver structure.

5. Stellate Cells: Approximately one-fifth of the cell volume within the sinusoids is occupied by hepatic stellate cells (HSC), which play a significant role in chronic toxic responses to substances like ethanol. These fat-storing cells, also known as Ito cells, reside in the small perisinusoidal gap between sinusoids and hepatocytes, referred to as the space of Disse (see Fig. 9.2). Stellate cells make up about 5% of the total cells in the liver and typically exist in a quiescent, or dormant, state. While in this inactive state, their physiological functions are not entirely clear but likely involve roles in storing vitamin A and other fat-soluble vitamins, as well as contributing to immune system functions. Additionally, stellate cells play a role in manufacturing the extracellular matrix, a crucial form of connective tissue that anchors liver cells in their positions within the overall organ structure.

Upon exposure to harmful chemicals, including alcohol, stellate cells undergo activation, transitioning into a highly efficient state dedicated to collagen production. The accumulation of collagenous scar tissue is a characteristic feature of liver fibrosis, a common consequence of chronic exposure to various hepatotoxicants, such as heavy metals, alcohol, and dietary mycotoxins.

6. Kupffer Cells: Another significant cell type within the liver is the Kupffer cell, a resident macrophage-like immune cell. The number of Kupffer cells in the liver typically surpasses resident phagocytic cell counts in other organs, and their distribution is higher in periportal zones compared to centrilobular areas. Equipped with toll-like surface receptors activated by ingested endotoxins, Kupffer cells play a crucial role in phagocytizing blood-borne toxicants and particulates received from the portal circulation.

Upon activation, Kupffer cells release a variety of mediators, including prostaglandins, cytokines, interferons, platelet-activating factors, lysosomal enzymes, and reactive oxygen species. In many toxic syndromes, these changes induce a persistent inflammatory state, subjecting neighboring tissue to continual exposure to damaging oxidants and inflammatory cytokines. This often leads to stellate cell activation and a prolonged fibrotic response to hepatotoxicants.

However, Kupffer cells are more than just defenders that combat invading pathogens or absorbed toxicants during stress. A substantial body of data suggests that they also play a role in various beneficial host defense pathways, indicating their versatility in contributing to both protective and potentially harmful responses within the liver.

9.4.1.1 Major toxic responses of the liver

Traditionally, chemicals causing liver injury were classified as intrinsic hepatotoxicants if they induced predictable, dose-related liver injury in most exposed individuals. In contrast, idiosyncratic hepatotoxicants typically triggered hepatotoxicity in a small minority (approximately <1 in 100,000) of exposed individuals. The severity of the hepatic response to an idiosyncratic hepatotoxicant was often assumed to have little relation to the administered dose, although recent studies suggest that this may not always be the case for drug-induced allergic hepatotoxicity. Distinguishing hepatotoxic responses can often be based on whether they can be reproduced in rodent species; intrinsic hepatotoxicants usually induce comparable hepatotoxicity in lab animals, whereas, idiosyncratic toxicity is often challenging to reproduce in rodents.

Historically, chemicals inducing intrinsic hepatotoxicity were assumed to cause liver injury via non-immunologic mechanisms, while idiosyncratic toxicity was typically attributed to immune-mediated responses, such as antibody production and T-cell activation. However, recent understandings challenge this assumption, suggesting that immunologic mechanisms contribute to the pathogenesis of many hepatotoxic syndromes, including acute intrinsic hepatotoxic syndromes once thought to exclude immunological pathways.

Studies on the effects of hepatotoxicants on liver histology have revealed a limited cluster of toxic responses. Therefore, toxic responses to a wide range of chemically diverse hepatotoxicants usually feature a few well-defined pathological responses.

One early sign of liver injury from diverse hepatotoxicants is steatosis, a reversible condition where small or large fatty vesicles (micro- or macro-vesicular droplets) appear within the cytoplasm of parenchymal cells throughout the liver lobule. These droplets typically consist of triglyceride-filled droplets coated with phospholipids and specific lipid droplet-associated proteins, notably members of the PAT protein family. Ongoing exposure to hepatotoxicants that induce droplet formation often leads to more pronounced liver pathology. Diagnosing xenobiotic-associated steatosis is challenging for clinicians as fatty liver occurs in various common health disorders, including nonalcoholic fatty liver disease (NAFLD) in patients with obesity and metabolic syndrome, as well as steatotic conditions in patients with micronutrient deficiencies or chronic viral infections.

The mechanisms behind fatty droplet accumulation in the intoxicated liver are complex. While long-standing explanations attribute fat deposition to toxicant-induced shifts in the hepatocellular redox state favoring the accumulation of fatty acids, newer insights from microarray studies highlight changes in the expression of numerous proteins involved in hepatic lipid metabolism, including dysregulation of transcription factors controlling fatty acid synthesis (e.g., SREBP) and fatty acid oxidation (e.g., PPAR). Microarray studies investigating steatogenesis indicate a transcriptional shift toward an 'adipogenic' state, wherein the liver enhances its capacity for fatty acid synthesis while concurrently downregulating fatty acid oxidation and the secretion of very low-density lipoproteins (VLDLs).

(A) Cell Death

Cell death within the liver typically occurs through either necrosis or apoptosis.

1. Apoptotic Cell Death: Apoptotic cell death involves small clusters of hepatocytes and follows a tightly arranged sequence of molecular events. This process includes the controlled digestion of cellular components by cell death enzymes such as caspases.

2. Hepatic Necrosis: Heavier intoxication with hepatotoxicants can induce a more overt, uncontrolled form of cell death known as hepatic necrosis. This irreversible process involves the death of many hepatocytes and may manifest as 'focal' (localized to specific zones, as in centrilobular or periportal necrosis) or 'fulminant' (rapid onset, affecting most liver cells). While the early stages involve reversible cell swelling, events quickly progress to irreversible necrotic tissue destruction.

Typical signs of cell death by necrosis include:

- Loss of membrane integrity
- Swelling of mitochondria and other intracellular organelles
- ATP depletion
- Loss of calcium homeostasis secondary to calcium influx

The latter change activates calcium-dependent endonucleases, proteases, and phospholipases that begin digesting key cell components, leading to cytoskeletal derangement and cell blebbing. Cell lysis and prominent changes in the organization of nuclear DNA (e.g., pyknosis, karyorrhexis, and karyolysis) are also seen in necrotic liver.

Hepatic necrosis can result from intoxication with various noxious substances, including medicinal agents, natural substances, and synthetic chemicals. In industrialized nations, paracetamol intoxication is often the leading cause of hepatic necrosis in emergency department patients. Although the liver has a remarkable regenerative capacity that ensures rapid regrowth after acute intoxication with paracetamol and other hepatotoxicants, some types of hepatic necrosis trigger the formation of persistent scar tissue.

(B) Impaired Bile Flow

As the largest gland in the body, the liver plays a crucial role in eliminating diverse endogenous waste products, foreign xenobiotics, and their metabolites through bile secretion. Bile, produced by the liver, also contains cholesterol-derived detergent-like substances that aid in emulsifying ingested fats. However, certain hepatotoxic chemicals can disrupt these essential hepatic functions by inducing cholestasis, which is a partial or complete arrest of bile flow.

Cholestasis has immediate physiological consequences, as bile serves as a major elimination route for bilirubin and biliverdin—toxic chromogenic pigments formed during the degradation of heme-rich red blood cells. In normal liver function, bilirubin and biliverdin are actively exported into the tubular canalicular network, eventually draining into the bile duct. However, some hepatotoxic chemicals inhibit the membrane transporters responsible for exporting these pigments across canalicular membranes. This inhibition leads to impaired hepatic clearance of biliverdin and bilirubin, causing their deposition in blood and body tissues. The accumulation of these pigments in dermal layers can result in a yellow tinge to the patient's skin, a condition known as jaundice.

(C) Jaundice

Jaundice, characterized by a hyperbilirubinemia state, is not only associated with hepatotoxicity induced by certain drugs but is also common in newborn babies due to poor conjugative metabolism within neonatal livers. In some drug-related hepatotoxicity syndromes, such as those associated with heavy use of anabolic steroids, jaundice can be a prominent symptom. Unresolved jaundice can lead to a medical emergency, as the accumulation of unconjugated bilirubin in body tissues, particularly the brain, may elicit cellular injury. Therefore, managing and resolving cholestasis is crucial to preventing severe complications associated with impaired bile flow and jaundice.

(D) Liver Fibrosis

Continuous exposure to hepatotoxic substances often results in an excessive accumulation of extracellular matrix proteins, particularly collagen, leading to the development of liver fibrosis. The primary contributors to fibrosis are activated hepatic stellate cells (HSCs), with significant support from myofibroblasts, particularly in depositing collagen fibers within the canalicular tracts of the liver lobule. Fibroblasts originating from extrahepatic tissues, including bone marrow, may also contribute to the process of fibrogenesis. Considerable attention has been given to the molecular factors initiating fibrosis. Although a complex combination of cytokines and other mediators can activate HSCs, transforming them from a quiescent state to profibrogenic entities, platelet-derived growth factor (PDGF) released from activated Kupffer cells is likely a crucial factor. Additional mediators aid in extending the fibrogenic activity of HSCs by inhibiting their normal apoptotic death. In cases of alcoholic fibrosis, persistent alcohol consumption can lead to the progression to the cirrhotic phase of alcoholic liver disease—a terminal state characterized by uncontrolled fiber deposition and widespread hepatocellular death, resulting in the liver becoming a shrunken mass of dysfunctional tissue.

(E) Liver Cancer

Primary liver cancer is relatively rare in Western nations but more prevalent in African and Eastern Asian populations. Geographical variations in food supply contamination with fungal carcinogens likely contribute to these differences. Liver cancer originates from monoclonal tumor cell clusters, typically arising when cells acquire genetic changes that confer growth advantages to their descendants. Tumors can develop in bile duct or hepatic blood vessel cells, but hepatocytes are the most common origin. The prognosis for liver cancer is often grim due to its highly metastatic and aggressive nature, with advanced stages at the time of diagnosis. Chronic exposure to alcohol, toxic metals like arsenic, and industrial reagents such as vinyl chloride can lead to liver tumors.

9.4.1.2 Major hepatotoxicant classes

Many structurally diverse chemicals, including medicinal agents, occupational reagents, and hepatotoxic substances from plants and fungi, have been associated with liver injury in humans. Herbal remedies and dietary supplements contribute to the global burden of liver injury. Geographical differences exist in the relative importance of causative agents, with antibiotics, anticonvulsants, and psychotropic drugs being leading causes in Western societies, while in Asia, herbs and dietary supplements are prominent. Hepatotoxicity is a significant challenge in clinical medicine and the pharmaceutical industry, with drug-induced liver injury (DILI) occurring in various scenarios. Large doses, chronic exposure, and allergic reactions can lead to hepatotoxicity. The discovery of hepatotoxic properties during drug development often leads to the discontinuation of the compound. Hepatotoxicity concerns have resulted in the removal or limited use of many drugs in the market, especially when rare forms of DILI become apparent in large, genetically diverse populations.

Paracetamol

Paracetamol, also known as acetaminophen, is a drug with a complex profile, resembling the conflicting characters of "Dr. Jekyll and Mr. Hyde." Widely accessible due to its safety at low doses, it takes on a life-threatening aspect in cases of overdose. Its popularity grew post-World War II as an aspirin alternative, and for years, its hazardous side remained largely unknown to clinicians. In 1966, Thompson and Prescott's case report highlighted severe liver damage in a patient who overdosed on paracetamol, sparking awareness. Subsequent publications increased paracetamol's prevalence in suicidal poisonings, making it a leading cause of acute liver damage admissions to emergency departments globally.

Paracetamol's diverse effects stem from its metabolism (Fig. 9.3). As a phenolic compound, approximately 90% of a low dose undergoes conjugative metabolism, forming harmless metabolites. A small portion is excreted unmetabolized in urine. The final metabolite, constituting only 2–3% of a typical adult dose, forms through 2-electron oxidation by enzymes like CYP2E1 or 3A4, resulting in the highly reactive quinonimine metabolite, NAPQI (Fig. 9.3).

Under normal conditions, NAPQI undergoes detoxication via conjugation with glutathione in a GST-catalyzed reaction. However, during a paracetamol overdose, overwhelmed conjugative pathways, due to depleted cofactors PAPS and UDPGA, lead to increased CYP oxidation, generating dangerously high NAPQI levels. This depletes hepatic glutathione stores, rendering the hepatocellular proteome vulnerable to attack. The Target Protein Database indicates that NAPQI targets around 36 proteins in paracetamol-poisoned livers. Efforts to deduce the biochemical basis for paracetamol hepatotoxicity from these protein identities have had limited success. A contemporary understanding of paracetamol hepatotoxicity suggests the immune system's activation, triggered by widespread protein damage and glutathione loss, as the primary determinant of liver damage (Fig. 9.3). The extent and severity of hepatotoxicity likely depend on the dominance of anti-inflammatory or proinflammatory immune responses within the liver of an individual paracetamol-poisoned patient.

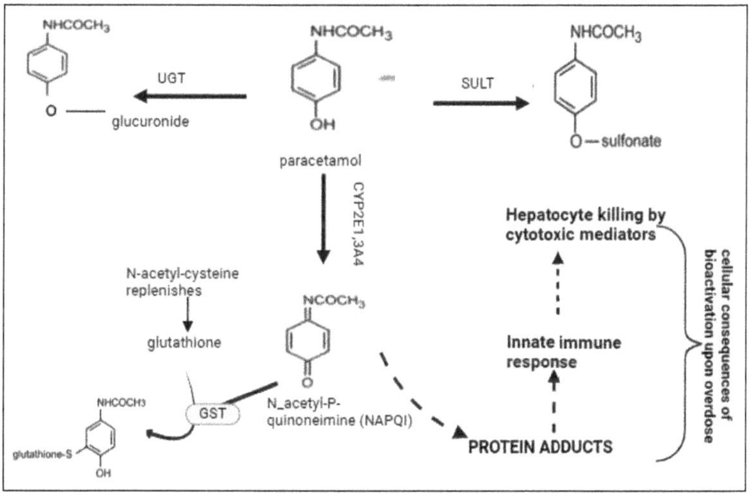

Figure 9.3 At normal doses, paracetamol is converted to nontoxic O-sulfonate and O-glucuronide metabolites, with only a small proportion of the drug undergoing oxidation to the toxic quinonimine NAPQI. These low levels of NAPQI are efficiently detoxicated via glutathione conjugation. Upon overdose, conjugative metabolism is overwhelmed and NAPQI causes extensive adduction of hepatocellular proteins, triggering liver cell death via a combination of pathways that include activation of the innate immune response.

Early recognition of NAPQI's role facilitated the identification of glutathione-replenishing drugs as antidotes for poisoned patients (Fig. 9.3). Susceptible mouse strains quickly became valuable models for testing various antidote drugs. Compounds mimicking glutathione effects within the liver, especially N-acetyl cysteine, demonstrated effectiveness in mice. However, N-acetyl cysteine, while generally effective, carries safety concerns, including a propensity for serious allergies in a small number of patients. Identifying patients at risk of hepatotoxicity in paracetamol poisoning remains challenging due to uncertainties about the actual dose ingested and the timing of ingestion. While determining paracetamol concentrations has been a mainstay in managing poisoned patients, these methods are not universally effective. The recent use of toxicogenomic approaches, such as microarrays, to monitor gene expression changes in blood cells from paracetamol-poisoned patients has shown promise. Specific clusters of mRNAs may serve as "signatures" to predict toxicity severity in individual patients. In an era of competition for hospital beds and rising healthcare costs, improvements in identifying paracetamol-poisoned patients at genuine risk of life-threatening outcomes would be highly beneficial.

Azidothymidine

In a significant accomplishment for modern pharmacology, within 25 years of identifying the human immunodeficiency virus (HIV) as the cause of AIDS, 25 antiretroviral drugs were developed to combat this devastating condition. The breakthrough in the mid-1990s, demonstrating the feasibility of suppressing AIDS by administering multiple anti-HIV drugs simultaneously, led to the introduction of highly active antiretroviral therapy (HAART) protocols. While the success in AIDS prevention was substantial, a drawback emerged as patients undergoing HAART, who consume a potent drug cocktail daily, exhibited an increased risk of drug-induced liver injury (DILI), making liver disease mortality a leading cause of death among HIV patients.

HAART-related hepatotoxicity is a complex and varied syndrome encompassing both acute and chronic disorders. Symptoms range from those typical of rare allergic DILI syndromes to conditions like hepatic steatosis with lactic acidosis and even manifestations resembling viral hepatitis, complicating the diagnosis. Clinicians managing long-term care for these patients face a challenging decision when encountering signs of nonspecific liver injury, such as asymptomatic elevation in blood transaminase levels. The dilemma arises as to whether a prescribing switch to a newer anti-HIV drug with lower hepatotoxic potential should be made when viral loads are effectively suppressed with the current medication.

Research into HAART-related hepatotoxicity has predominantly focused on the nucleotide/nucleotide reverse transcriptase inhibitor (NRTI) family of HIV drugs. Since the discovery of the first member of this class, azidothymidine (AZT), in 1985, NRTIs have been a cornerstone of anti-HIV therapy. NRTIs inhibit reverse transcriptase (RT), a crucial viral protein necessary for HIV reproduction, belonging to the atypical RNA-dependent family of DNA polymerases. Despite their success in suppressing HIV and NRTIs, particularly AZT, they have concerns about hepatotoxicity, prompting investigations into the mechanisms and potential alternatives within this drug class. Nucleotide reverse transcriptase inhibitors (NRTIs) typically mimic nucleotides and undergo phosphorylation at their 5'-hydroxy group by cellular kinases before being incorporated into DNA by reverse transcriptase (RT). The absence of a free 3'-hydroxy group prevents further DNA synthesis, effectively terminating the process. Despite the success of this strategy in suppressing HIV in infected patients, the clinical use of NRTIs comes with various limitations, including the emergence of drug-resistant RT mutants and host toxicity issues such as neuropathy. Hepatotoxicity is a particular concern among these limitations.

The main cause of NRTI-induced hepatotoxicity is the inhibition of human mitochondrial DNA polymerase γ (Pol γ). Pol γ, belonging to the A family of DNA polymerases, structurally resembles HIV RT. This structural similarity makes Pol γ more likely than other human DNA polymerases to utilize NRTIs as enzyme substrates during DNA replication. Since Pol γ plays a crucial role in copying the mitochondrial genome, inhibiting this enzyme with NRTIs leads to side effects consistent with a loss of mitochondrial function in the liver of drug-treated patients. This impairment results in deficient ATP production and renders the liver vulnerable to mitochondrial uncoupling, leading to the overproduction of damaging free radicals. While the liver is the primary target for such toxicity, other tissues may also exhibit mitochondrial pathology during prolonged azidothymidine (AZT) therapy. The recognition of Pol γ inhibition in NRTI side effects has spurred the development of alternative NRTIs with fewer associated problems. However, the cost-effectiveness of AZT relative to newer drugs still contributes to its popularity in some countries, particularly in the developing world where HIV infection is a significant concern.

Troglitazone

Troglitazone, approved in several countries in 1997, was hailed as a significant breakthrough in treating type 2 diabetes, a metabolic disorder characterized by inadequate control of blood sugar levels. Representing the first member of the new thiazolidinedione class of antidiabetic medications, troglitazone acted as an "insulin sensitizer" by binding to PPARγ, a member of the peroxisome proliferator-activated receptor family. This action aimed to enhance the removal of glucose from the blood by body tissues, particularly after meals.

Despite its promising properties, concerns arose shortly after troglitazone's market release. Doctors in Japan, Britain, and the USA reported cases of life-threatening idiosyncratic hepatotoxicity. Subsequently, the manufacturers withdrew the drug from the market a few years later due to safety issues. The adverse effects led to approximately 60–70 deaths worldwide being attributed to troglitazone.

Troglitazone-induced liver disease, like many idiosyncratic toxicities, was challenging to predict, with serious hepatotoxic injury estimated to occur in approximately 1 in 10,000–20,000 patients. The mechanisms underlying troglitazone hepatotoxicity are likely complex. The parent drug may contribute since it can induce apoptosis in both hepatocytes and non-hepatic cells. However, metabolism and bioactivation are also likely factors, as the unique vitamin E–like chroman ring of troglitazone undergoes CYP-catalyzed oxidation to various toxic intermediates, including ring-opened electrophiles, quinone metabolites, phenoxy radicals, and epoxides.

Given that troglitazone is a ligand for the xenosensor protein PXR, a potent inducer of CYP3A expression, it is probable that troglitazone increases the formation of reactive metabolites by inducing its CYP-dependent bioactivation. Japanese researchers studying genetic polymorphisms in xenobiotic biotransformation pathways as risk factors in patients who developed troglitazone hepatotoxicity found that patients carrying twin null mutations in two key glutathione-S-transferase genes (GSTT1 and GSTM1) were significantly overrepresented among at-risk patients. These mutations likely compromise the cellular ability to detoxify electrophilic metabolites, strengthening the proposed role of reactive metabolites in the pathogenesis of at least some cases of troglitazone hepatotoxicity. Ongoing research aims to further understand the mechanisms involved in troglitazone hepatotoxicity, as insights into identifying at-risk patients for rare idiosyncratic toxic responses can enhance our understanding of other drug-related syndromes.

9.4.2 Nephrotoxicity

Chemical damage to the kidneys, known as chemically induced nephrotoxicity, is closely linked to damage in the liver, called hepatotoxicity. Both these organs can be harmed by the same toxic substances at the same time. For instance, exposure to cadmium, a heavy metal found in batteries, paints, and plastics, can cause damage to both the liver and kidneys. The extent of damage depends on how much cadmium a person is exposed to and for how long. Large, sudden doses of cadmium typically harm the liver, while extended exposure to lower doses can affect the kidneys. Unlike the liver, the kidneys cannot regenerate, so episodes of kidney damage may lead to long-term issues, such as the need for dialysis or a kidney transplant.

There are several reasons why the kidneys are prone to chemical toxicity. The kidneys have a high concentration of transporters that move toxic substances, making them vulnerable to damage because the concentration of these substances can be much higher in the kidneys than in the bloodstream. The large surface area of the kidneys and the presence of enzymes that metabolize toxic substances in certain kidney structures, like the proximal tubules, further increase the risk. The kidneys' blood vessels are also sensitive to compounds that affect blood flow, making the kidneys more susceptible to injury because changes in blood flow can increase the concentration of toxic substances in the kidney tissue. Due to these factors, exposure to substances that are toxic to the kidneys is a significant concern in clinical practice. Xenobiotic (foreign substances) poisonings contribute to about half of all cases of sudden and ongoing kidney failure. Additionally, 10–15% of admissions to intensive care units involve sudden kidney failure. Despite advances in medicine, the death rates for sudden kidney failure have not significantly decreased in the last 50 years, emphasizing the continued need for research in this field.

The kidneys, like the liver, play a crucial role in maintaining the overall well-being of various physiological systems. They contribute significantly to the balance of electrolytes in the body by regulating the volume and ionic composition of the total bodily fluid, including water, sodium, potassium, and hydrogen ions. Additionally, the kidneys produce essential hormones that control blood pressure (such as renin), stimulate red blood cell production (erythropoietin), and regulate blood calcium levels (calcitriol). The ability of the kidneys to eliminate foreign substances and metabolic waste products (like urea, ammonia, and uric acid) is vital for overall bodily health.

1. Basic Renal Structure: Every day, the adult human kidneys filter around 180 liters of blood—an impressive accomplishment for two small organs about the size of a typical computer mouse. Each kidney is protected by a renal capsule, a clear membrane pouch guarding against infections and trauma. When viewed in the transverse section, each kidney consists of an outer renal cortex and an inner renal medulla. A single renal artery originating in the aorta supplies blood to each kidney, and blood exits the organ through a renal vein that drains into the inferior vena cava. The urine produced by each kidney is transported to the bladder through a single ureter.

2. The Renal Nephron: The nephron is the fundamental functional unit of the kidney, a lengthy tubular structure responsible for filtering the blood and eliminating bodily wastes. Each kidney consists of around one million nephrons, categorized into various subclasses based on the location of their glomerular apparatus, such as mid-cortical and juxtamedullary. This leads to significant diversity in the spatial distribution of nephron subtypes and the morphology of epithelial cells within each nephron region. The specific impact of individual nephrotoxicants on distinct renal functions reflects the anatomical complexity of the organ, ensuring that different chemicals tend to target specific segments of the renal nephron. The tendency for chemicals to accumulate in different kidney regions also influences nephrotoxic syndromes.

3. The Glomerulus: The glomerulus serves as the primary filtering unit of the kidney, acting as the initial point where circulating blood interacts with the filtered biofluid, eventually forming urine. Each glomerulus is supplied with blood through afferent arterioles and drained by efferent arterioles, consisting of three main components: endothelial cells, the glomerular basement membrane, and visceral epithelial cells. As blood enters the glomerulus, it undergoes processing to create an ultrafiltrate in Bowman's space, containing water and small solutes. In healthy individuals, blood cells and large negatively charged proteins do not enter the ultrafiltrate due to the intact glomerular membrane.

Renal ultrafiltration is facilitated by hydrostatic pressure within the glomerulus, influenced by the relative vasoconstriction of the afferent and efferent arterioles. Various factors, such as sympathetic nerves and mediators like endothelin, angiotensin II, vasopressin, prostaglandins, and cytokines, control renal vascular tone. The glomerular filtration rate (GFR), representing the extent of filtration at the glomerulus, remains constant in healthy adults. It can be estimated by monitoring the excretion of model compounds that are filtered but not reabsorbed or actively excreted (e.g., inulin). Blood concentrations of endogenous molecules like creatinine, urea, or cystatin C, typically cleared by glomerular filtration, can be indicators of renal toxicity in clinical settings. Elevated plasma levels of these substances suggest renal dysfunction.

Because blood-borne xenobiotics (foreign substances) are delivered to the glomerulus in high concentrations, exposure to various nephrotoxicants may lead to glomerular damage (Fig. 9.4). Unfortunately, confirming glomerular injury through renal biopsies is rarely done in emergency room settings, potentially resulting in the underestimation of its contribution to acute human nephrotoxicity. Apart from acute injury, prolonged exposure to environmental pollutants like heavy metals and hydrocarbons can trigger glomerulonephritis. This condition involves a significant reduction in blood-filtering capacity and urine formation. Reactive oxygen species and cytokines released by neutrophils and macrophages trapped within the inflamed glomerulus are likely mediators of this immune-mediated renal injury.

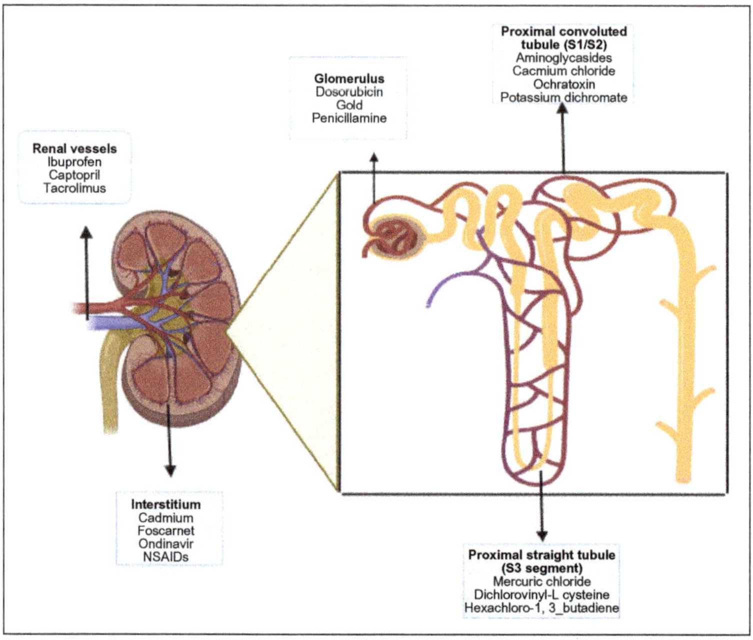

Figure 9.4 Nephrotoxicants often display selectivity when causing kidney injury by either damaging specific regions within the renal nephron (insert), causing vasoconstriction within specific renal zones, or inducing prolonged inflammatory damage within the interstitium (interstitial nephritis).

4. Proximal Tubules: The proximal tubule consists of specialized epithelial cells with distinct features, including a basolateral membrane zone interacting with blood vessels and a highly invaginated membrane known as the brush border. These tubules are crucial for reabsorbing about two-thirds of solutes and water from the ultrafiltrate, aided by aquaporin-1 water channels. Proximal tubules are of significant toxicological importance because most renal CYP expression is located there. These structures also strongly express multidrug resistance proteins (MRP1–5) and various transport proteins, making them susceptible to nephrotoxicity due to their high capacity for accumulating circulating toxicants.

The proximal tubules are vulnerable to toxic metabolites that can induce cell rupture, releasing protein aggregates that promote tubular obstruction. Some nephrotoxicants, like the immunosuppressant cyclosporine, can impair urine production by causing interstitial nephritis, a chronic condition involving inflammation in the space between renal tubules.

The proximal tubules consist of three distinct segments (S1, S2, and S3). S1 epithelial cells are highly metabolically active, containing many mitochondria and a vacuolar lysozyme system to clear small proteins from the ultrafiltrate. S1 cells have an extensive capacity for active secretion of xenobiotics, making them highly vulnerable to nephrotoxicants. S2 cells possess similar properties but with more limited metabolic and excretory capacities compared to S1 cells. S3 cells are cuboidal with long brush borders and high numbers of peroxisomes.

5. Loop of Henle, Distal Tubules, and Collecting Duct: The terminal segments of the renal nephron consist of several structurally and functionally distinct parts. The descending and ascending arms of the loop of Henle have clear differences in water permeability, with the descending arm showing a high expression of aquaporins for water permeability and the ascending arm exhibiting low aquaporin expression. The distal tubules have limited water permeability but are extensively involved in transporting sodium ions. The collecting duct plays a crucial role in determining the final composition of urine by hormonally regulating the reabsorption of solutes, water, and hydrogen ions. Unlike the glomerulus or proximal tubules, the distal tubules, and collecting duct have limited expression of CYP, making chemically induced toxicity less common in these regions.

6. Functional Consequences of Nephrotoxicity: As a general rule, the functional changes associated with toxicant-induced nephrotoxicity can either be acute, developing rapidly following initial exposure to a harmful drug or chemical, or progressive, developing slowly and insidiously during chronic exposure to nephrotoxic substances. In some cases, long-term exposure to certain toxic substances may lead to renal cancer.

Functional consequences of xenobiotic-induced renal injury often manifest through changes in the quantity or quality of urine. Symptoms may involve the overproduction of dilute urine (polyuria), excretion of low quantities of urine (anuria), or the passage of blood-stained urine (hematuria). Changes in the composition of urine, such as the presence of glucose (glucosuria) or blood proteins (proteinuria), can indicate renal damage. Specific proteins like albumin (albuminuria) are useful indicators of nephrotoxicity in a clinical setting. Unfortunately, renal injury is often well advanced before such changes become obvious to patients, emphasizing the need for early markers of drug- or toxicant-induced nephrotoxicity.

Researchers have explored specific proteins within urine as markers of injury to distinct renal structures. Traditional biochemical approaches were initially used, but their clinical adoption was limited. In recent years, broad-based metabolomic and proteomic methods for monitoring numerous molecules in urine have identified many promising nephrotoxicity biomarkers. While these markers are still undergoing validation in animal studies and human trials, they hold significant potential for clinical use in the future.

9.4.2.1 Major human nephrotoxicants

The development of a unified mechanistic understanding of chemically induced nephrotoxicity has faced challenges due to the dissimilarities between nephrotoxicants in terms of their chemical structures and physicochemical properties. Despite clinical signs of kidney damage often being similar, there are few chemical or structural similarities among diverse toxicants, ranging from heavy metals like mercury and mycotoxins such as fumonisin B1 to immunosuppressants like cyclosporine A and antibiotics like the aminoglycoside gentamicin. Instead of attempting to cover a comprehensive range of xenobiotic-induced nephrotoxic syndromes, we will focus on several examples that highlight different mechanisms underlying the toxic response.

Nephrotoxic Drugs

A small number of human pharmaceuticals are overrepresented in drug-related human kidney failure. Two commonly involved drug classes deserve particular attention.

1. Aminoglycosides: Despite significant concerns about renal injury and hearing loss, aminoglycoside antibiotics like gentamicin, tobramycin, and amikacin remain in modern use due to their effectiveness against serious Gram-negative bacterial infections. These drugs, being highly ionized polycations, are typically administered through intravenous injections as they are poorly absorbed orally. Even with monitored plasma concentrations within the therapeutic range, a significant number of patients may still experience renal injury. Aminoglycoside nephrotoxicity is attributed to the efficient active transport of these drugs by renal transporters in proximal tubules, while some members of this drug class may be less nephrotoxic than others. The risk of renal injury is likely higher in the elderly and in patients taking multiple drugs. In laboratory animals, aminoglycoside nephrotoxicity manifests as progressive epithelial cell necrosis in the S1 and S2 segments of proximal tubules. As the renal injury progresses, tubular cell mitochondria undergo swelling and distortion, accompanied by structural abnormalities such as cristae and matrix loss. Notably, lysosomal abnormalities and the formation of myeloid bodies are also observed. Although the molecular mechanisms underlying aminoglycoside nephrotoxicity are not fully clear, they likely involve interference with ion and nutrient transport at the apical and basolateral membranes, along with inhibitory effects on mammalian protein synthesis at the high concentrations reached in proximal tubular cells.

2. NSAIDs: Nonsteroidal anti-inflammatory drugs (NSAIDs) are widely used, with common members like aspirin, ibuprofen, and naproxen available over the counter. NSAIDs are effective inhibitors of cyclooxygenase, enzymes involved in pain and inflammation in various conditions. Cyclooxygenases produce eicosanoids, including prostaglandins and prostacyclins, crucial for renal functions like glomerular filtration, water/ion transport, and blood vessel tone. While NSAID-induced renal toxicity is rare in healthy individuals, patients with pre-existing renal impairment (e.g., elderly) face a risk of serious nephrotoxicity. NSAIDs, by suppressing cyclooxygenase-derived eicosanoids, can lead to a sudden decrease in urine output, signaling acute renal failure in some patients. Besides these predictable consequences, chronic NSAID use can result in other less predictable renal toxicities. Long-term aspirin use may lead to acute tubular necrosis, and extensive NSAID use can occasionally cause rare interstitial nephritis, highlighting the potential serious nephrotoxic effects of these drugs.

Prevention

Prevention of nephrotoxicity involves careful consideration and monitoring of potentially nephrotoxic agents, particularly medications, by healthcare providers. This includes selecting medications with lower nephrotoxic potential when feasible, prescribing appropriate dosages, and closely monitoring patients' kidney function during treatment, especially in those with pre-existing kidney disease or other risk factors. Patient education is also essential to ensure individuals are aware of the potential risks, adhere to prescribed medication regimens, and avoid unnecessary exposure to environmental toxins. Regularly assessing and addressing modifiable risk factors, such as hypertension and diabetes, can further reduce the risk of kidney injury. Additionally, healthcare professionals should remain vigilant in recognizing early signs of nephrotoxicity and take prompt action to mitigate further damage when it occurs.

9.4.3 Neurotoxicity

1. Introduction to Neurotoxicity: Neurotoxic chemicals play a significant role in the health effects resulting from environmental and workplace exposures. The National Institute for Occupational Safety and Health (NIOSH) identifies exposure to neurotoxic chemicals as a leading cause of work-related diseases and injuries. Over 25% of chemicals with established Threshold Limit Values (TLV) by the American Conference of Governmental Industrial Hygienists (ACGIH) demonstrate nervous system effects, and approximately 42% of commonly used chemicals are considered neurotoxic.

Studying the toxic effects on the nervous system is challenging due to the complexity of the human nervous system, which distinguishes us from other organisms. The nervous system's complexity results in diverse responses to toxicant exposure, affecting physiological, behavioral, cognitive, and emotional aspects. Quantifying these effects is challenging, and their significance for human health may not always be clear. For instance, changes in mood or temporary decreases in reaction time may not be life-threatening on their own, but in an industrial setting with multiple hazards, they can lead to disasters.

Not all industrial chemicals are neurotoxic, but for those that are, neurotoxic effects often serve as sensitive indicators of low-level exposure. Monitoring these subtle effects requires the development of appropriate methods. Changes in behavior are commonly used as sensitive neurotoxic endpoints, though quantifying them objectively and dealing with baseline variability among individuals can be challenging. The chapter later describes some standard tests for neurotoxicity.

The nervous system, known for its complexity, is widespread throughout the body. It can be divided into two main parts: the central nervous system, primarily consisting of the brain and spinal cord, and the peripheral nervous system, which includes sensory and motor nerves, among other components. This division is significant in the context of neurotoxicity because certain neurotoxicants seem to affect either the central or peripheral nervous systems exclusively.

2. Mechanisms of Neuronal Transmission: The nervous system is a vast network of interconnected nerve cells, or neurons, supported by various auxiliary cell types. While neurons vary in shapes, sizes, and functions, they generally consist of dendrites, a cell body, and an axon. Dendrites receive chemical signals from adjacent neurons, initiating electrical impulses along the axon and prompting the release of more chemical signals at the terminal boutons. This process allows a stimulus to travel throughout the entire human body.

To maintain the electrical impulse, neurons often rely on the myelin sheath, acting as an insulator around the axon. Neurons meet at synapses, gaps where chemical signals (neurotransmitters) diffuse from one neuron to the dendrites of the next. Neurons may also terminate at muscles or glands, releasing neurotransmitters to specialized receptors at these sites.

These fundamental features of the nervous system are similar across a wide range of organisms, instilling confidence in using neurotoxicity test results in animals to predict potential neurotoxic effects in humans.

3. The Action Potential: Electrical signals along the axon are generated through a phenomenon known as an action potential. This potential arises from a charge difference across the nerve membrane, influenced by the movement of sodium (Na^+), potassium

(K^+), and chloride (Cl^-) ions. The membrane's selective permeability, concentration and potential gradients, and active transport contribute to this charge difference. In the resting state, K^+ ions have a higher concentration inside the cell, while Na^+ and Cl^- ions have higher concentrations outside, maintaining balance through a potential gradient.

In this equilibrium, the membrane is relatively impermeable to Na^+, resulting in a net positive charge outside the cell compared to the inside. When the cell is stimulated and an action potential occurs, the membrane becomes permeable to sodium, causing an influx of positive charge and depolarization of the membrane. This depolarization is propagated down the axon, with current flowing ahead of the action potential. Behind the action potential, the membrane's permeability shifts to favor K^+ movement, leading to repolarization as K^+ ions move out of the cell and restore the membrane to its original resting potential.

4. Neurotransmitter Activity: Communication between neurons and other cells involves both electrical and chemical signals. Electrical signals are the fastest means of communication and occur through gap junctions, and membrane pores connecting tightly packed neurons. Chemical signals involve neurotransmitters released at the synapse, binding to receptors on the postsynaptic cell and triggering a response. Neurotransmitters, released from vesicles in response to the action potential in the presynaptic cell, play a crucial role in this process.

The most common neurotransmitter in the human body is acetylcholine (ACh), primarily found in the peripheral nervous system. ACh usually acts in an excitatory manner, initiating an action potential in the postsynaptic neuron, but it can also inhibit signal propagation at certain synapses. Acetylcholinesterase (AChE) hydrolyzes ACh after it triggers a response, terminating its activity. Future signals rely on newly synthesized ACh molecules. Failure to hydrolyze older molecules, as seen with some neurotoxicants, can lead to inappropriate and continuous stimulation of the postsynaptic cell by accumulated ACh.

Various neurotransmitters exist, including γ-aminobutyric acid (GABA), amines (epinephrine, norepinephrine, dopamine, serotonin), amino acids (glycine, glutamate), and peptides (enkephalins, endorphins). These neurotransmitters may have excitatory or inhibitory actions and target neurons, muscle fibers, or glandular cells.

9.4.3.1 Agents that act on the neuron

Membrane Disruption

The effective transmission of neuronal signals is contingent on the integrity of membranes along the axon's length and at the terminus where neurotransmitters are released. Many neurons are equipped with myelin, which aids in propagating action potentials by minimizing potential loss across the membrane. Chemicals that disrupt this membrane system can lead to impairments in nervous system function.

Certain industrial solvents, such as methanol, trichloroethylene, and tetrachloroethylene, commonly employed for cleaning purposes due to their lipophilicity, have the potential to harm cell membrane lipids. Inhalation exposure to these volatile solvents poses a significant threat to the nervous system.

Metals, like lead, thallium, and triethyltin, encountered during industrial processes such as smelting, can disrupt the myelin sheath in both the central and peripheral nervous systems. The extent of damage varies, ranging from vision loss associated with thallium exposure to impaired cognition linked to lead exposure.

Insecticides, including the banned DDT and others, interfere with membrane function by disrupting ion channels responsible for maintaining sodium and potassium balance. For example, DDT inhibits repolarization after action potential propagation, leading to symptoms such as tremors and seizures. Despite the ban on DDT, organochlorine insecticides and pyrethroid esters, which act similarly, are still in use worldwide, posing potential hazards for workers involved in manufacturing and application processes.

Peripheral Sensory and Motor Nerves

Peripheral sensorimotor nerve effects result in abnormal sensation and impaired motor control in distal body regions, particularly hands and feet. Peripheral nerves can regenerate if axonal damage is limited, allowing complete reversal with prompt toxicant exposure limitation. This contrasts with CNS neuropathy, which usually causes permanent damage.

Certain peripheral neurotoxicants, like solvents methyl n-butyl ketone and n-hexane, mainly exert effects through liver-formed toxic metabolites. Distal axonal swelling in motor and sensory neurons, primarily caused by the metabolite 2,5-hexanedione, is observed. Carbon disulfide, another solvent, induces a similar effect, but the role of its metabolites is less clear.

Some chemicals transform outside the body into products closely associated with the parent compound, confusing the actual cause. Trichloroethylene inhalation, known to target the central nervous system, has long been associated with cranial nerve effects. Recent studies suggest cranial neuropathy attributed to trichloroethylene inhalation may be caused by dichloroacetylene, a breakdown product occurring in some industrial settings. Dichloroacetylene has been shown to target cranial nerves; whereas, this has not been demonstrated with pure trichloroethylene.

Permanent Brain Lesions

Damage to neurons in the brain can result in a range of outcomes depending on the affected area, from impaired sensory, cognitive, or motor skills to severe or fatal consequences. Lesions in the brain are often irreversible due to the high metabolic rate of neuronal tissue. The brain, being highly perfused with blood vessels, has evolved the blood-brain barrier (BBB) to protect against neurotoxic chemicals in the bloodstream.

The BBB consists of tightly joined cells with few transport vesicles, reducing membrane permeability to many blood-borne chemicals. Although effective against most large or hydrophilic molecules, some highly lipophilic molecules can traverse the BBB. An example is the neurotoxic metal mercury. In its ionized or inorganic salt form, mercury is water-soluble and less likely to cross the BBB. However, the neurotoxic symptoms of mercury poisoning, including tremors, mood disorders, psychosis, and potentially death, manifest when lipophilic elemental or organic mercury species are formed.

In the brain, elemental or organic mercury, which can more easily cross membranes, may undergo transformation into an ionized form, leading to prolonged exposure and severe brain lesions. This transformation from inorganic to organic mercury can occur in the environment, often facilitated by bacteria, or within the human gut through bacterial activity.

9.4.3.2 General population exposure to environmental neurotoxicants

Neurotoxicants, in addition to being workplace hazards, can enter the environment through intentional or unintentional releases, leading to exposure for the general population through air, food, or drinking water. An infamous example is the Minamata incident in the 1950s, where residents of a Japanese town experienced severe neurotoxicity from methylmercury poisoning due to contaminated fish.

Identifying a specific neurotoxicant causing a particular effect in the face of various environmental chemicals can be challenging. Environmental exposure levels are typically not high enough to cause immediate symptoms in everyone, often affecting only certain individuals, presumably those most highly exposed or sensitive. Neurotoxic effects may manifest as changes in behavior, serving as early warnings of potential toxicity if exposure continues.

Dissolved metals like lead and arsenic in drinking water or organic solvents such as trichloroethylene and carbon tetrachloride in groundwater and surface water can lead to neurotoxic effects. While exposure levels are often deemed not high enough to pose serious risks, the impact of low-level, chronic exposure on developing nervous systems remains a concern, especially for fetuses and infants.

Inhalation exposure is generally considered less of a threat outside industrial settings, as volatile chemicals disperse easily in the atmosphere. People near industrial air pollution sources might face elevated levels of neurotoxic chemicals. Concerns also arise in confined spaces, such as inhaling volatile organic compounds from contaminated water during showering or exposure to elevated levels of chemicals from dry-cleaned clothes in enclosed spaces like houses or cars. However, the correlation with neurotoxic effects is not fully understood.

9.4.4 Cardiotoxicity

9.4.4.1 General considerations

The cardiovascular system comprises the heart and blood vessels, playing a crucial role in distributing essential energy sources to various organs. Although the heart is not a common target for xenobiotic-related toxicity, it can be adversely affected by a range of chemicals. These substances may exert their impact either directly on the myocardium or indirectly through the nervous system or blood vessels. Smooth muscle cells and endothelial cells within blood vessels are particularly susceptible to chemical-mediated toxicity, as they are consistently exposed to absorbed xenobiotics through systemic circulation. Table 9.1 enumerates chemicals possessing vascular toxic potential.

Table 9.1 Vasculotoxic drugs and chemicals

Smooth muscle dysfunction	
Arsenic	Heavy metals
Sorafenib	Anticancer drugs
5-Fluorouracil	Anticancer drugs
Endothelial impairment	
Rofecoxib	Anti-inflammatory drugs
Monocrotaline	Plant toxin
Cisplatin	Anticancer drugs
VEGF inhibitors	Anticancer drugs

9.4.4.2 Toxic effects on the heart

The heart primarily consists of myocardial cells, each approximately measuring 15 μm × 80 μm. In contrast to skeletal muscle cells, not all heart muscle cells are innervated; however, those that are connected to each other through nexus points, facilitate rapid electrical stimulus transmission. This connectivity is crucial for the coordinated contraction sequence of different heart regions. The myocardium differs from skeletal muscles in having less contractile material (50% vs. 80%) but more mitochondrial matter (35% vs. 2%). Mitochondria play a vital role in cardiac contractility and serve as a common subcellular target for cardiotoxicity.

Myocardial contraction entails the liberation of energy via oxidative metabolism, its preservation through adenosine triphosphate and creatine phosphate, and its deployment by contractile proteins. Fragile mechanisms encompass energy consumption and intracellular movement of calcium ions (crucial for myocardial contractility, enzyme regulation, and hormonal information transduction). Disruptions in these signaling pathways at both cellular and organ levels result in cardiotoxicity. The rising apprehension revolves around drug-induced cardiotoxicity, evident in chemotherapy-associated cardiac dysfunction underscoring the imperative for heightened surveillance and comprehension of the mode of action to avert chemical-induced cardiotoxicity.

9.4.4.3 Cardiomyopathy

The primary mechanisms underlying drug- and chemical-induced cardiomyopathy involve dysregulation of intracellular calcium, mitochondrial impairments, and oxidative stress. Cobalt has been implicated in numerous severe and fatal cases of cardiomyopathy. Studies suggest that cobalt-associated cardiotoxicity may be mediated by the intramitochondrial accumulation of calcium. The toxic effects of cobalt on the heart are significantly heightened in cases of malnutrition, particularly the deficiency of specific amino acids. Research by Grice (1972) indicated that cobalt ions decrease oxygen uptake and disrupt cardiac energy metabolism in the tricarboxylic acid cycle, particularly under thiamine deficiency. Cobalt has the potential to decrease available myocardial Ca^{2+} by forming complexes with macromolecules and may also act antagonistically to endogenous Ca^{2+}.

Adrenergic β-receptor agonists, specifically isoproterenol, and vasodilating antihypertensive drugs like hydralazine, have the potential to induce myocardial necrosis. The former exerts direct adrenergic effects, while the latter induces adrenergic effects through the resulting hypotension. These effects lead to an increased transmembrane calcium influx, subsequently enhancing the rate and force of contraction. This, coupled with concurrent hypotension, results in cardiac hypoxia. The combination of hypoxia and calcium deposits in the mitochondria contributes to the breakdown of organelles and sarcolemma.

Anthracycline antibiotics, including doxorubicin, are known to be associated with acute adverse effects such as hypotension, tachycardia, and arrhythmias. Prolonged administration of these drugs leads to degeneration and atrophy of cardiac muscle cells, interstitial edema, and fibrosis. Anthracycline antibiotics are recognized for generating reactive oxygen species (ROS) in cardiac myocytes, resulting in oxidative damage to proteins, nucleic acids, and lipids. These modifications to key macromolecules in cardiomyocytes ultimately disrupt contractile function and lead to cell death through apoptosis or necrosis.

The mechanism of doxorubicin-induced cardiotoxicity is complex, involving apoptosis mediated by calcium overload and iron-catalyzed formation of free radicals. Despite its anticancer mode of action attributed to the inhibition of topoisomerase II (Bernard et al., 2011), doxorubicin, with quinones in its chemical structure, enhances ROS generation in primary cultures of cardiomyocytes in a time- and concentration-dependent manner.

9.4.4.4 Interference with nucleic acid synthesis and DNA damage

As mentioned earlier, the induction of oxidative stress is proposed as a primary mechanism for the cardiotoxicity of anthracycline antibiotics such as doxorubicin and daunorubicin. Additionally, the binding of these antibiotics to mitochondrial and nuclear DNA, leading to interference with RNA and protein synthesis, represents a potential mechanism. This impact on the heart is significant due to the relatively short half-life of contractile proteins (12 weeks). Other conceivable mechanisms include the peroxidation of membrane lipids and hypotension resulting from the release of cytokines. It is noteworthy that mitochondrial DNA is highly vulnerable to oxidative stress compared to nuclear DNA, given that mitochondria are a primary cellular site for the generation of reactive oxygen species (ROS), and protective mechanisms against DNA damage, such as histones or a DNA repair system, are limited in mitochondria. Oxidative damage to mitochondrial DNA may contribute to cardiotoxicity, as observed in conditions like alcoholic cardiomyopathy.

9.4.4.5 Arrhythmias

Several fluorocarbons have the potential to induce cardiac arrhythmias. This effect is brought about by sensitizing the heart to epinephrine, depressing contractility, reducing coronary blood flow, and triggering a reflex increase in sympathetic and vagal impulses to the heart after irritating the mucosa in the respiratory tract.

Cardiac arrhythmias can also be induced by tricyclic antidepressants, likely resulting from imbalances within the autonomic regulatory system of the heart. Propylene glycol, a commonly used solvent, can convert ventricular tachycardia induced by deslanoside into ventricular fibrillation.

Taxanes, a class of antineoplastic chemotherapeutic agents, are implicated in inducing arrhythmias as one of their most common toxicities. Paclitaxel, a specific taxane, has been shown to decrease calcium increase and contraction in cardiomyocytes, along with reducing the time from the maximum contracted state to relaxation. Various arrhythmias, including bradyarrhythmias and supraventricular arrhythmias such as atrial fibrillation, atrial flutter, and atrial tachycardias, were observed following exposure to taxanes.

9.4.4.6 Myocardial depression

Several lipid-soluble organic compounds, including general anesthetics, have the capability to depress cardiac contractility. The likely mechanism involves a nonspecific expansion of various cellular membranes through the insertion of chemically indifferent molecules into the hydrophobic regions of integral proteins and membrane phospholipids.

Antibiotics, such as amphotericin B, chloramphenicol, streptomycin, and tetracycline, can induce hypotension by depressing cardiac contractility. The mechanism of action is believed to be associated with the inhibition of Ca^{2+} bound to superficial membrane sites.

9.4.4.7 Toxic effects on blood vessels

The vascular system comprises arteries, arterioles, capillaries, venules, and veins. Toxicants have the potential to impact any of these vessels, and the severity of the effect is contingent on the physiological role of the organ supplied by the affected blood vessels. Figure 9.5 illustrates representative adverse symptoms associated with vascular toxicity. Vascular smooth muscle cells and endothelial cells are susceptible to various Vaso toxicants, and the mechanism of vascular toxicity varies depending on the target molecules affected by the toxicants. Representative mechanisms include the heightened generation of reactive oxygen species (ROS), disruption of the cGMP/NO/NOS pathway, or DNA damage for endothelial toxicity. In the case of smooth muscle toxicity, mechanisms involve the interruption of the contractile machinery, such as the Rho kinase pathway, phosphorylation of the myosin light chain, and calcium dysregulation.

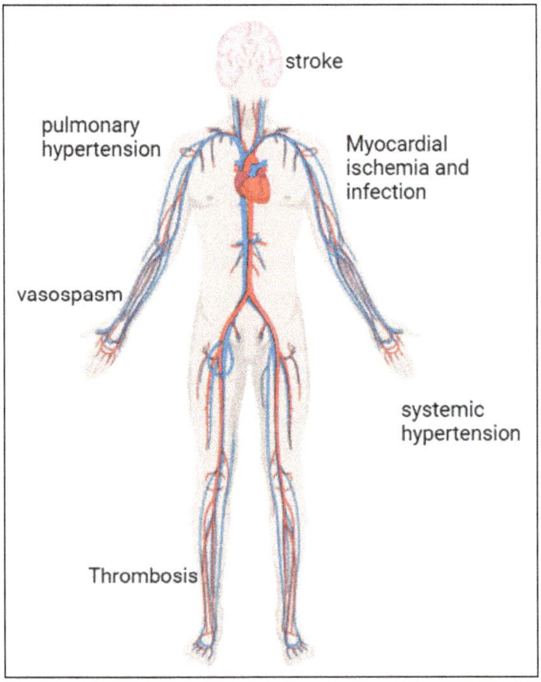

Figure 9.5 Vascular toxicity and related symptoms.

9.4.4.8 Vasoconstriction and vasodilation

Ingesting ergot alkaloids, which are fungal contaminants found in specific foods, can result in gangrene due to vasoconstriction. Prolonged exposure to pure oxygen, especially in premature infants, may lead to blindness, attributed to associated vasoconstriction in the eyes. Arsenic is known to induce calcium sensitization, leading to increased vasoconstriction in smooth muscle cells. The anticancer agent sorafenib has the potential to cause impaired regulation of Rho kinase, contributing to possible vasospasm.

Occupational workers exposed to nitroglycerin have been reported to face sudden mortality due to a heart attack on their days off. It seems that continued exposure to a coronary dilator accustomed the workers to a low level of coronary flow, and the abrupt cessation of exposure to the coronary dilator precipitated coronary insufficiency.

Large doses of specific hypotensive drugs, such as minoxidil and hydralazine, have the potential to damage the musculature of the coronary artery. This lesion may result from exaggerated pharmacodynamic changes.

9.4.4.9 Fibrosis

Cadmium and lead have the potential to affect blood vessels in the kidneys, leading to renal fibrosis. This can result in the impairment of blood supply, which may interfere with the "non-excretory functions" of the kidneys, indirectly contributing to the development of hypertension.

9.4.4.10 Tumors

Certain toxicants have been associated with the development of tumors in blood vessels. For instance, exposure to vinyl chloride has been reported to induce hemangiosarcoma in the liver of both humans and animals. Additionally, hemangioendothelioma has been linked to exposure to thorium dioxide.

9.4.5 Reproductive Toxicity

Reproductive toxicity is the occurrence of adverse effects on male or female sexual anatomy, function, maturation, or behavior resulting from exposure to exogenous chemical agents. This definition includes lactation, sexual maturation, the ability to produce viable, fertile offspring, sex-specific behavior, and premature reproductive senescence. There is now speculation that gender preference and gender identity may also be influenced by toxic effects. Because the potential for successful reproduction begins in the womb, it can be difficult to distinguish reproductive toxicity from developmental toxicity.

9.4.5.1 Disruption of male reproduction by toxicants

The male reproductive physiology is susceptible to disruption by toxicants at various stages of development and function, including gestation. Toxic effects to the testes can lead to diverse effects, such as a reduction in sperm count and blood androgen levels that fall either above or below the normal range. Damage to the epididymis can result in poor sperm motility or inadequate semen quality. Since sperm production is directly linked to the number of Sertoli cells and cannot be replenished once lost, failure to properly differentiate this crucial cell population can permanently limit the capacity to produce viable sperm. Disruption to the testis or accessory sex glands can occur through various mechanisms, including DNA damage, inhibition of cell division, interference with GnRH secretion, or alteration of testicular vasculature. The most extensively studied and best-understood mechanism of action is interference with androgen receptors (ARs) and/or androgen production.

Pesticides

One of the earliest animal studies designed to test the hypothesis that chemical agents could interfere with androgen action was conducted in 1950 using chickens. Injection of the pesticide dichlorodiphenyltrichloroethane (DDT) resulted in markedly undersized testes and inhibited the development of the comb and wattle. It was later determined that DDT and its metabolites function as anti-androgens and compete with endogenous androgens for access to the ARs. Because androgens, especially DHT, are essential for the development of male genitalia, it is widely speculated that gestational exposure to compounds like DDT with anti-androgen properties may be responsible for the documented increase in the frequency of developmental defects in male genitalia including hypospadias (misplaced urethral opening on the penis) and cryptorchidism (undescended testes) in boys. Impaired fertility from pesticide exposure has also been documented in humans. For example, one of the earliest known examples of reduced male fertility in humans following pesticide exposure emerged in the 1970s in a group of 25 Californian men exposed occupationally to the pesticide 1,2-dibromo-3-chloropropane (DBCP). DBCP was a potent nematicide that went into commercial production in 1956 and was later banned in 1977. The exact mechanism by which DBCP impairs male reproduction has still not been elucidated, but it is hypothesized that one of its metabolites acts directly on germ cells, thereby impairing spermatogenesis. Of the 25 occupationally exposed men, most had reduced sperm counts and 9 had no viable sperm. Subsequent studies, which examined the long-term effects of exposure on these and other occupationally exposed men, found that sperm production ultimately improved in most cases, but not in individuals whose sperm counts were lowest.

These studies illustrate two important concepts. First, effects on the male reproductive system from exposure to pesticides and other toxicants during adulthood are largely temporary and reversible because the system is already developed and, in most cases, can recover. Second, recovery is unlikely if primary spermatogonia or Sertoli cells are killed. In the case of DBCP, long-term exposure depleted the pool of healthy spermatogonia, ultimately resulting in irreversible infertility.

Metals

Among metals, the impact of lead on the male reproductive system is extensively documented and understood. Lead exposure has been associated with reduced sperm counts and an increased frequency of malformed sperm, leading to diminished fertility. The effects of lead on the testis are diverse. It can directly interfere with the hypothalamic-pituitary-gonadal (HPG) axis, resulting in hypogonadism or low testosterone levels. Additionally, lead can disrupt the formation and function of the vascular system within the testis, ultimately causing damage to the seminiferous tubules. Lead exposure is most common in industrialized settings but can also occur in older homes through the inhalation or ingestion of lead paint residues. Children are particularly vulnerable to exposure through this route or consuming water with unacceptably high lead levels. While there is some concern that exposure to lead levels considered "safe" may affect brain development and impair male fertility, the data on this remains inconclusive.

Cadmium is another metal well-known for its detrimental effects on male reproduction, but typically only at near-fatal doses. It induces Sertoli cell death, testicular necrosis, and damage to germinal cells through a direct impact on the vascular system within the testis. Exposures of this magnitude are more likely to occur in industrial settings. Mercury and boron are also metals suspected of having reproductive effects in men, although research on these is still ongoing.

Plastics

Considerable attention is now being placed on the potential for chemicals with endocrine-disrupting properties to leach from plastics into food. Of these, the most widely studied group that pertains to male reproductive health is the phthalates. There are many kinds of phthalates, and the two considered to have the greatest potential to impact male reproduction are dibutyl phthalate (DBP) and diethylhexyl phthalate (DEHP). DBP is used in many personal care products such as lotions, cosmetics, nail polish, and perfume. It is also found in the coatings of many time-release pharmaceuticals. DEHP is primarily used as a plasticizer in the production of flexible products including vinyl, medical tubing, and other medical devices, baby toys, and flooring. A series of studies conducted in rats in the late 1990s was the first to demonstrate that phthalates could interfere with the ability of testosterone to masculinize the male reproductive tract. Exposure in utero, when the genitals are being formed and the action of DHT is critical, resulted in several genital malformations in laboratory animals including hypospadias and hemorrhagic testes. Interestingly, the phthalates do not produce their effects by antagonizing ARs, but rather by interfering with the production of testosterone in the fetal testis. Exposure to phthalates during pregnancy has now been associated with a greater risk of genital malformations in infant boys. Epidemiological evidence has also positively correlated higher urinary phthalate levels with lower sperm counts and an increased likelihood of sperm with damaged DNA in adult men. Phthalate exposure is also linked with lower circulating levels of estrogens and androgens, indicating that the HPG axis may be impaired. Although it is important to keep in mind that correlation does not prove causation, these newly emerging epidemiology studies are the best evidence to date that phthalates have the potential to affect male reproductive health.

9.4.5.2 Disruption of female reproduction by toxicants

Data on female reproductive health trends are limited. However, recent data suggest that conception rates are declining in the developing world and the prevalence of reproductive disorders such as advanced pubertal onset, premature ovarian failure, and polycystic ovary syndrome are becoming more common. Delayed childbearing resulting from cultural changes arguably has the biggest impact on fertility trends in Western countries. Obesity, dietary habits, and chronic stress may also be important factors affecting female reproductive health. The reported increase in infertility, particularly among women under the age of 25, cannot be fully explained by conventional factors. It is now widely hypothesized that environmental exposure, whether to the mother or her developing fetus, may also play a role. Female reproductive physiology is susceptible to disruption by toxicants at various stages of development and function, including gestation. Toxic effects on the ovaries can lead to structural abnormalities or damage to the follicles. Compounds acting as androgen agonists could potentially masculinize aspects of the female reproductive tract or brain. For instance, exposure to low levels of androgens during fetal life is believed to increase the risk of developing polycystic ovary syndrome.

Disruption of uterine development can occur through various mechanisms, including endocrine disruption, inhibition of cell division, or improper maturation of the Müllerian ducts. Improper ductal differentiation could result in malformations within the fallopian tubes, potentially compromising the ability to sustain pregnancy. The most extensively studied and best-understood mechanism of action is interference with estrogen receptors and/or estrogen production. This interference can have profound effects on the female reproductive system and may contribute to the reported increase in infertility, especially among younger women.

Cigarette Smoke

Smoking has long been known to induce premature menopause and reduce fertility. Smokers have fewer eggs, fewer quality eggs, an impaired capacity to recruit quality eggs when administered fertility drugs, and decreased rates of successful in vitro fertilization. Cigarette smoke contains several hundred known toxicants, all of which could, either individually or in combination, produce these effects. In animal models, nicotine has been shown to have disruptive effects on ovulation rates, oocyte maturation, and implantation success. Chromosomal abnormalities within oocytes are also more prevalent in smokers compared to the general population and may contribute to lower conception rates among smokers. Fertility in the offspring of smokers may also be compromised. Studies in mice have found that in-utero exposure to polycyclic aromatic hydrocarbons, a group of more than 100 different chemicals found in cigarette smoke, results in significantly fewer ovarian follicles and therefore compromised reproductive capacity.

Diethylstilbestrol (DES)

One of the most well-understood examples illustrating how the administration of synthetic estrogen can impact female reproduction is the tragic case of diethylstilbestrol (DES), which was used by pregnant women in the mid-20th century. Introduced in 1938, DES, a potent synthetic estrogen, was prescribed to women for the prevention of miscarriage until 1971. It is estimated that 4–6 million women were prescribed DES in the United States alone during this period. The consequences of in-utero DES exposure first came to light in 1971 when a group of vigilant physicians observed that DES daughters were more likely to develop an extremely rare type of cervicovaginal cancer.

Since the initial discovery, other abnormalities in DES daughters have been identified, including uterine malformations, an increased risk of ectopic pregnancy, premature menopause, a higher risk of developing uterine fibroids and endometriosis, an increased risk of breast cancer, and decreased fertility. There is also emerging evidence that the children of DES daughters, referred to as DES granddaughters, are experiencing reproductive problems. For these women, exposure occurred when they were germ cells in their mother's developing ovary in the womb of their grandmother. This case stands as one of the earliest instances in humans demonstrating that persistent, generational effects can result from in-utero exposure to a potent estrogen.

Pesticides

A variety of pesticides and herbicides have been found to interfere with hormone action, especially estrogen. The most famous of these is DDT (still sold under many names including Anofex, Dicophane, Dinocide, Neocid, and Neocidol, among others). DDT is usually sold and used as a mixture of several, closely related compounds of which dichlorodiphenyldichloroethylene (DDE) is a component and the principal breakdown product. Created in 1874 by Othmar Zeidler, the insecticidal properties were not revealed until 1939 by Hermann Muller, for which he was awarded the 1948 Nobel Prize in Physiology and Medicine. This insecticide, which is structurally like DES, was widely used all over the globe until the early 1970s to eradicate lice and mosquitoes. It was, and continues to be, a critical weapon in the fight against malaria and other mosquito-borne illnesses which, collectively, cause the greatest number of preventable deaths in the world. More people die from malaria than HIV, heart disease, or cancer. For that reason, it is still used in many places where the risk of contracting malaria remains high although some mosquito populations have developed resistance to the insecticide. The effects of DDT on wildlife populations were popularized by the landmark 1962 book Silent Spring by biologist Rachel Carson. The book became an instant bestseller and launched the environmental movement in the United States as well as a governmental investigation into its claims that persistent pesticides, including DDT, were poisoning wildlife and, potentially, humans. DDT and its metabolites are toxic to a wide range of animals and insects. It thins eggshells in birds and is believed to be one of the primary culprits for the decline of the bald eagle and other large birds of prey. DDT has multiple modes of action. It is moderately toxic to mammals with a rat oral lethal dose 50 (LD50) of 113 mg/kg and is like the pyrethroids in that it is a nervous system stimulant that acts by opening neuronal sodium channels, causing spasms, seizures, and death. At far lower doses, it is also thought to be both an estrogen agonist and an androgen antagonist. Whether or not DDT can impair reproductive development in women is still the subject of investigation. Exposure in utero or early in life is associated with an increased risk of breast cancer (Cohn et al., 2007). There is also emerging evidence from areas where DDT is still used, that DDT exposure is associated with preterm birth, early pregnancy loss, reduced semen quality, disrupted menstruation, and problems with lactation. Infants from these regions are frequently exposed to levels that exceed the acceptable daily intake established by the World Health Organization via breast milk. Therefore, the potential reproductive risk of DDT exposure is ongoing for many populations outside the United States.

Plastics

A plethora of chemicals are employed in plastics manufacturing to enhance their utility, with additions aimed at reducing microbial growth or improving stability, clarity, durability, or pliability. One additive that has garnered substantial scientific and political attention is bisphenol A (BPA). Initially developed in the 1930s as a synthetic estrogen, BPA was abandoned as a pharmaceutical with the invention of DES. However, it is now widely used in various products, notably polycarbonate plastics, to enhance strength and durability. BPA is also found in dental sealants, epoxy resins, and the linings of food cans and drink cartons, including soda cans. The annual worldwide production of BPA is estimated to exceed 6 billion pounds; and in 2008, the U.S. Center for Disease Control (CDC) estimated that BPA was present in approximately 95% of the U.S. population.

A study from early 2009 revealed that exposure to BPA, along with phthalates and other endocrine disruptors leaching from plastics, is particularly high among premature infants in neonatal intensive care units. Although BPA does not persist in the environment like DDT or PCBs, daily human exposure is estimated to be higher due to its ready leaching from plastic containers, especially when heated. Laboratory rodents exposed to BPA in utero or just after birth exhibit various reproductive abnormalities, including accelerated puberty, abnormal ovarian follicles, premature loss of the ovulatory cycle, and alterations in mammary gland formation, suggesting an elevated risk of developing breast cancer.

Chapter 10

Air Pollution

"The silent threat in the air we breathe"

10.1 Introduction

The environment refers to the surroundings and can be comprehensively defined as the total of all conditions and influences affecting the development and life of organisms. Life on the Earth exists within a specific zone called the biosphere, which supports life. The biosphere can be divided into the atmosphere (air), lithosphere (soil), and hydrosphere (water), based on the life-supporting media.

Since gaining independence, India has made significant progress and is now among the top 10 industrialized nations globally. However, development often comes with environmental degradation. The environment is continuously burdened with foreign chemicals (xenobiotics) released by urban communities and industries. Various chemicals are introduced into different parts of the environment through industrial, agricultural, public health operations, and automobile emissions. Due to the lack of an effective environmental management or pollution control system, many chemical substances remain in the environment. These chemicals may either exceed safe levels or stay below threshold limits. Currently, around 9 million chemicals are known worldwide, with approximately 30,000 new

Environmental Toxicology: Concepts and Principles
Mohd Salim Reshi, Darakhshan Javaid, Syed Sanober Qadri, and Neelu Sinha

ISBN 978-981-5129-68-7 (Hardcover), 978-1-003-63707-3 (eBook)
www.jennystanford.com

chemicals added annually. Of these, 75,000–80,000 are commonly used in daily life, agriculture, and public health. While some chemicals improve productivity, living standards, and health, many are potentially hazardous.

Pollution is any undesirable change in the physical, chemical, and biological characteristics of air, land, and water, which can negatively impact human, animal, and plant life. Environmental pollution refers to the contamination of the environment with impurities, making it unfit for its intended use. Any substance causing pollution is known as a pollutant, and when pollutant levels become toxic to life, they are termed toxicants. When pollutants reach lethal levels in the environment, they become environmental toxicants. The study of these toxicants, including their nature, sources, environmental fate, and harmful effects on living organisms, is known as *entox* or *Environmental Toxicology*. This field focuses on understanding the effects of pollutants on the structure and function of ecosystems.

Air pollution is the presence of undesirable solid or gaseous particles in the air in quantities that are harmful to human health and the environment. A substance in the air that can be adverse to humans and the environment is known as an air pollutant. Pollutants can be in the form of solid particles, liquid droplets, or gases. In addition, they may be natural or man-made. Pollutants can be classified as primary or secondary. Usually, primary pollutants are directly produced from a process, such as ash from a volcanic eruption, carbon monoxide gas from a motor vehicle exhaust, or sulfur dioxide released from factories. Secondary pollutants are not emitted directly. Rather, they form in the air when primary pollutants react or interact. An important example of a secondary pollutant is ground-level ozone – one of the many secondary pollutants that make up photochemical smog. Some pollutants may be both primary and secondary, that is, they are both emitted directly and formed from other primary pollutants.

10.2 Importance of Air Quality for Human Health and the Environment

Air quality stands as a critical determinant of both human well-being and environmental equilibrium, underscoring its paramount importance in contemporary society. The quality of the air we breathe directly influences human health, with profound implications for respiratory and cardiovascular systems. Exposure to air pollutants, such as particulate matter, ground-level ozone, and nitrogen oxides, has been linked to a spectrum of adverse health effects, ranging from aggravated asthma to increased risk of heart disease and lung cancer. Furthermore, vulnerable populations, including children, the elderly, and individuals with pre-existing health conditions, face heightened susceptibility to the deleterious impacts of poor air quality. Beyond its direct impact on human health, air pollution bears grave consequences for the environment. Ecosystems and biodiversity are threatened as pollutants contribute to soil and water contamination, while acid rain, a byproduct of air pollution, exacerbates ecological imbalances. The air quality and climate change, mediated by greenhouse gas emissions, further highlight the global significance of maintaining pristine air quality. Addressing the challenges of air pollution emerges not only as a health imperative but as an ethical responsibility to safeguard the delicate balance of our shared environment.

10.3 Historical Context and Evolution of Air Pollution Concerns

The historical context and evolution of air pollution concerns trace a narrative that reflects humanity's growing awareness of the environmental consequences of industrialization and technological progress. In the early stages of the Industrial Revolution, as factories proliferated and urban centers expanded, the emission of pollutants into the air escalated significantly. However, the adverse effects of this pollution often went unnoticed or were thought inevitable consequence of industrial growth. It was not until the mid-20th century that a paradigm shift occurred, propelled by seminal events that forced society to confront the tangible impacts of air pollution. The infamous London Smog of 1952, a lethal combination of fog and high sulfur dioxide levels, resulted in thousands of deaths and served as a wake-up call, prompting the British government to enact the Clean Air Act in 1956. Concurrently, in the United States, incidents such as the Donora Smog of 1948 and the Cuyahoga River catching fire in 1969 brought public attention to the environmental perils associated with unchecked industrial emissions. These events marked pivotal moments in the evolution of air pollution concerns, catalyzing legislative initiatives and the establishment of regulatory frameworks globally. The subsequent decades witnessed an intensification of scientific research, international cooperation, and public advocacy, culminating in landmark agreements like the 1970 Clean Air Act in the United States and the formation of the Environmental Protection Agency (EPA). As air pollution continued to transcend national boundaries, the need for concerted global efforts became evident, leading to international agreements such as the Montreal Protocol and the Kyoto Protocol. The historical development of air pollution concerns highlights a shared understanding of the link between human activities and environmental health, emphasizing the need to maintain a balance for the benefit of current and future generations.

10.4 Sources of Air Pollution

The following are the common sources of air pollution (Fig. 10.1):

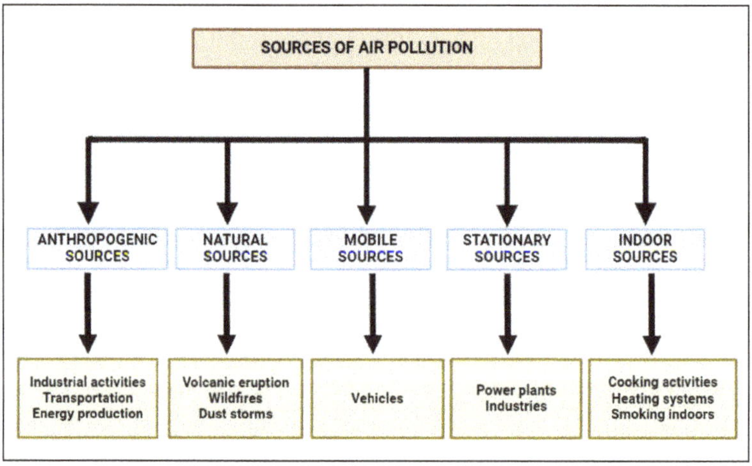

Figure 10.1 Common sources of air pollution.

1. Anthropogenic Sources: Human activities are major contributors to air pollution, releasing a variety of pollutants into the atmosphere. Industrial activities, such as manufacturing processes and the combustion of fossil fuels, emit pollutants like sulfur dioxide (SO_2), nitrogen oxides (NOx), particulate matter, and volatile organic compounds (VOCs). Transportation, driven by cars, trucks, and airplanes, releases significant amounts of carbon monoxide (CO), nitrogen oxides, and particulate matter. Energy production, particularly from power plants, is another key anthropogenic source, emitting pollutants such as carbon dioxide (CO_2), sulfur dioxide (SO_2), and nitrogen oxides.

2. Natural Sources: While human activities are prominent sources of air pollution, nature also contributes to the composition of the atmosphere. Volcanic eruptions release gases like sulfur dioxide and particulate matter into the air, influencing air quality on a regional scale. Wildfires, a natural occurrence, emit particulate matter, carbon monoxide, and other pollutants. Dust storms, driven by natural processes such as wind erosion, transport large amounts of particulate matter over long distances, affecting air quality in affected regions.

3. Mobile Sources: Mobile sources encompass vehicles that move, and their combustion engines emit pollutants. Cars, trucks, and airplanes release pollutants like carbon monoxide, nitrogen oxides, hydrocarbons, and particulate matter. The widespread use of these vehicles for transportation contributes significantly to urban air pollution and global emissions.

4. Stationary Sources: Stationary sources are fixed emitters of pollutants and include power plants and industrial facilities. Power plants burning fossil fuels release greenhouse gases (CO_2) and air pollutants like sulfur dioxide and nitrogen oxides. Industries emit a variety of pollutants depending on their processes, contributing to both local and global air quality concerns.

5. Indoor Sources: Indoor sources of air pollution are often overlooked, yet they play a crucial role in human exposure to harmful substances. Cooking activities, especially when using solid fuels, can release indoor air pollutants such as particulate matter and carbon monoxide. Heating systems, particularly those using wood or coal, contribute to indoor air pollution. Smoking indoors releases numerous harmful chemicals, adversely affecting air quality and posing health risks to occupants.

10.5 Major Air Pollutants

Some major air pollutants are discussed here (Fig. 10.2).

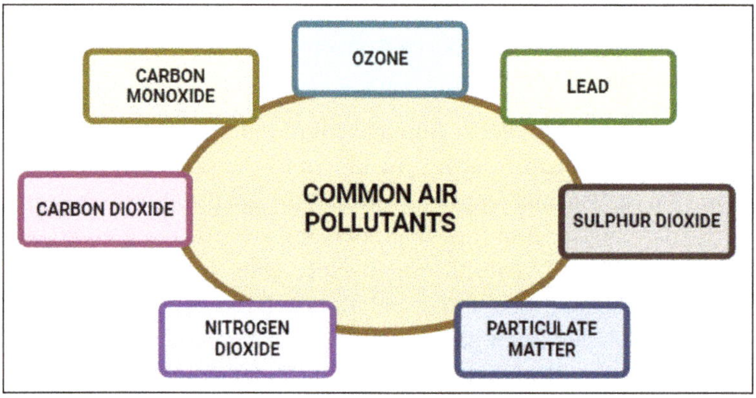

Figure 10.2 Some common air pollutants present in the atmosphere.

There are five primary pollutants that together contribute about 90% of the global air pollution. These are carbon oxides (CO and CO_2), nitrogen oxides, sulfur oxides, volatile organic compounds (mostly hydrocarbons), and suspended particulate matter.

1. Carbon Dioxide: Carbon dioxide is one of the major gases that contribute to air pollution. It is mainly produced during the combustion of fuel in factories, power stations, households, etc. The increasing CO_2 in the atmosphere is likely to have the following effects:

(i) A rise in atmospheric temperature or global warming due to the greenhouse effect causes climate change.

(ii) Reduced productivity of the marine ecosystem is because water in the oceans would be more acidic due to increased concentration of CO_2 in the air, which dissolves in the water.

(iii) Due to global warming, the increased surface temperature would cause the melting of continental and mountain glaciers and thus would cause flooding of coastal areas of some countries.

2. Sulfur Dioxide: Sulfur oxides are produced when sulfur-containing fossil fuels are burnt and coal in powerhouses and automobiles (cars, trucks, etc.) are burnt. It causes chlorosis and necrosis of plants, irritation in the eyes, and injury to the respiratory tract (asthma, bronchitis) in humans responsible for discoloration and deterioration of buildings. High concentration of sulfur dioxide in the atmosphere dissolves in raindrops to form sulfuric acid, which causes acid rain.

3. Carbon Monoxide: Carbon monoxide is produced because of incomplete combustion of fossil fuels like coal, petroleum, and wood charcoal. Automobiles using diesel and petrol are the major sources of carbon monoxide, which gets added to the atmosphere. Carbon monoxide is more dangerous than carbon dioxide. It is a poisonous gas that causes respiratory problems. When it reaches the bloodstream, it replaces oxygen due to its high affinity for hemoglobin. It also causes giddiness and headache and interferes with the normal function of the heart.

4. Lead: Lead is a metal found naturally in the environment and in manufactured products. It is one of six criteria pollutants for which EPA has established protective standards. Exposure to lead can occur through multiple pathways, including inhalation of air and ingestion of lead in food, water, soil, or dust. Historically, the major sources of lead emissions were motor vehicles (such as cars and trucks) and industrial sources. Due to the phase-out of leaded gasoline, however, airborne lead is no longer a problem in most of the developed and developing world. Excessive lead exposure can cause seizures, brain and kidney damage, mental retardation, and behavioral disorders. Children who are 6 years of age and under are most at risk because their bodies are growing quickly.

5. Fluorides: Upon heating, rocks, soils, and minerals that contain fluorides, give out hydrogen fluoride gas. This is an extremely toxic gas, which causes serious injury to livestock and cattle. It causes irritation, diseases of bone (fluorosis), mottling of teeth, respiratory diseases in humans, and destruction of crops.

6. Oxides of Nitrogen: A few oxides of nitrogen, such as nitric oxide (NO), nitrous oxide (N_2O), and nitrogen dioxide (NO_2) are produced by natural processes as well as from thermal power stations, factories, automobiles, and aircraft (due to burning of coal and petroleum). Nitrogen dioxide is a brownish, highly reactive gas present in all urban atmospheres. The major mechanism for the formation of nitrogen dioxide in the atmosphere is the oxidation of the primary air pollutant nitric oxide. They reduce the oxygen-carrying capacity of the blood and may cause eye irritation, bronchitis, edema of the lungs, and skin cancer in human beings.

Health effects of exposure to nitrogen dioxide include the following: In children and adults with respiratory diseases such as asthma, nitrogen dioxide can cause coughing, wheezing, and shortness of breath. Even short exposures to nitrogen dioxide can affect lung function. In children, short-term exposure can increase the risk of respiratory illness.

7. Particulate Matter: Particulate matter (PM) is the term for small particles found in the air including dust, dirt, soot, smoke, and liquid droplets. Sources of PM include cars, trucks, buses, factories, construction sites, tilled fields, unpaved roads, construction, wood burning, agricultural burning, wildfires, prescribed fires, and natural windblown dust.

Particulate matter is a criteria pollutant that comes in a wide range of sizes.

(i) Particles less than 10 micrometers in diameter tend to pose the greatest health concern because they can be inhaled into and accumulate in the respiratory system.

(ii) Particles less than 2.5 micrometers in diameter are referred to as fine particles.

(iii) Sources of fine particles include all types of combustion (e.g., motor vehicles, power plants, and wood burning) and some industrial processes.

(iv) Particles with diameters between 2.5 and 10 micrometers are referred to as coarse. Sources of coarse particles include crushing or grinding.

Particulate matter (or aerosols), either in solid or liquid form, is categorized based on the aerodynamic diameter. Particles less than 100 microns are also called "inhalable" since they can easily enter the nose and mouth. Particles less than 10 microns (PM10, often labeled "fine" in Europe). These particles are also called "thoracic" since they can penetrate deep into the respiratory system.

Particles less than 4 microns. These particles are often called "respirable" because they are small enough to pass completely through the respiratory system and enter the bloodstream.

8. Pollens: Pollens are the natural source of air pollution that can significantly affect air quality and health. Produced by plants during their reproductive cycles, pollen grains are tiny, lightweight, and can be carried over long distances by the wind. While essential for plant reproduction, pollen can also be a major allergen for many people. Different types of plants release pollen at various times of the year, leading to seasonal allergies. For instance, tree pollen from species like oak, birch, maple, and poplar is common in the spring. Poplar trees produce large amounts of pollen that can contribute to significant allergy symptoms. Grass pollen, such as from timothy, ryegrass, and bluegrass, tends to be prevalent in the late spring and early summer. Ragweed pollen, a notorious allergen, peaks in late summer and early fall. Exposure to high levels of pollen can trigger allergic reactions, including sneezing, runny nose, itchy eyes, and even asthma attacks in sensitive individuals. Understanding the patterns and effects of pollen as a natural air pollutant is crucial for developing effective strategies to mitigate its impact on public health.

10.5.1 Domestic Air Pollutants

Smoke from cigarettes, cigars, and other such objects from burning of tobacco, coal, firewood, cow dung cakes, kerosene oil, and liquefied gases are major domestic pollutants. The common pollutant gases emitted during the domestic burning of coal, kerosene oil, firewood, cow dung cakes, etc. are carbon monoxide (CO), carbon dioxide

(CO_2), sulfur dioxide (SO_2), etc. The pollution, due to these pollutants, causes suffocation, eye and lung diseases, and low visibility.

Table 10.1 Pollutants and their effect on humans and plants

Pollutants	Effects on man, vegetation, and other materials
Carcinogenic hydrocarbon	Cancer
Carbon monoxide	Poisoning, increased accident liability
Dust	Respiratory diseases, diseases like silicosis (cough, cold, sneezing, allergic diseases, etc.), asbestosis, byssinosis, poisoning from metallic dust
Hydrogen sulfide	Irritation of respiratory passages, danger of respiratory paralysis and asphyxiation On materials – Darkening of painted surfaces, corrosion
Hydrogen fluoride	Irritation, diseases of bone (fluorosis), mottling of teeth, respiratory diseases On vegetation – destruction of crops
Heavy metals	Specific poisoning, retardation of activities of the brain, interference in enzyme activities in the liver and kidney
Nitrogen dioxide	Irritation, bronchitis, edema of the lungs
Photochemical smog (oxidants)	Irritation, bronchitis, edema of the lungs
Sulfur dioxide	Suffocation, irritation of throat and eyes, respiratory diseases On vegetation – destruction of sensitive crops and reduced yield On materials – corrosion

10.6 Health Impacts of Exposure

Both fine and coarse particles can accumulate in the respiratory system and are associated with numerous health effects.

(i) Coarse particles can aggravate respiratory conditions such as asthma

(ii) Exposure to fine particles is associated with several serious health effects, including premature death. Adverse health effects have been associated with exposures to PM over both short periods (such as a day) and longer periods (a year or more).

(iii) When exposed to PM, people with existing heart or lung diseases, such as asthma, chronic obstructive pulmonary disease, congestive heart disease, and ischemic heart disease, are at increased risk of premature death or ailments.

(iv) When exposed to PM, children and people with existing lung disease may not be able to breathe normally and may experience symptoms such as coughing and shortness of breath.

10.7 Environmental Impacts

10.7.1 Effects on Ecosystems

Air pollution poses a significant threat to ecosystems, exerting detrimental effects on biodiversity, ecological balance, and overall environmental health. One of the most pressing consequences of air pollution is the alteration of atmospheric composition, leading to the release of harmful pollutants such as particulate matter, nitrogen oxides, sulfur dioxide, and volatile organic compounds. These pollutants, often emitted from industries, vehicular exhaust, and other anthropogenic sources, can have profound and widespread impacts on both terrestrial and aquatic ecosystems.

In terrestrial ecosystems, the deposition of pollutants from the atmosphere can lead to soil acidification, posing a threat to plant life. Acid rain, a consequence of elevated sulfur dioxide and nitrogen oxide emissions, can alter the pH of soils, making them inhospitable for many plant species. This disrupts nutrient cycling and availability, affecting the composition of plant communities and, consequently, the entire food web. Furthermore, air pollutants can directly damage plant tissues, impair photosynthesis, and compromise the overall health and vitality of vegetation. As a result, the cascading effects extend to herbivores and omnivores dependent on these plants, impacting population dynamics and community structure.

Aquatic ecosystems, too, bear the brunt of air pollution. Deposition of pollutants into water bodies through atmospheric pathways, a process known as atmospheric deposition, can lead to acidification of lakes and rivers. This acidification not only threatens aquatic flora and fauna directly but also alters the availability of essential nutrients, affecting the entire aquatic food web. Additionally, airborne pollutants can contribute to the formation of harmful algal blooms, further disrupting aquatic ecosystems. The toxic substances released into the air, such as heavy metals, can accumulate in water bodies, posing a threat to aquatic life and potentially entering the food chain with far-reaching consequences.

Beyond direct impacts on flora and fauna, air pollution can trigger complex ecological interactions and feedback loops. For example, the decline of certain plant species due to air pollution can disrupt habitat structures, affecting the availability of resources for other organisms. This, in turn, can lead to shifts in species composition and abundance, potentially resulting in the dominance of more pollution-tolerant species. Such changes in ecosystem structure and function may compromise the delivery of ecosystem services, including water purification, pollination, and climate regulation.

Moreover, air pollution has implications for global biodiversity, as pollutants can be transported over long distances, affecting ecosystems far from their original sources. This underscores the interconnectedness of ecosystems on a global scale and emphasizes the need for coordinated international efforts to address the root causes of air pollution.

10.8 Greenhouse Gases and Their Role in Climate Change

Greenhouse gases (GHGs) are crucial components of the Earth's atmosphere, playing a natural role in regulating the planet's temperature. However, human activities have significantly increased the concentration of these gases, leading to enhanced heat-trapping effects and contributing to global climate change. Key greenhouse gases include carbon dioxide (CO_2), methane (CH_4), nitrous oxide (N_2O), fluorinated gases, and water vapor.

1. Carbon Dioxide (CO_2): Carbon dioxide is a greenhouse gas primarily emitted through human activities, notably the combustion of fossil fuels such as coal, oil, and natural gas. Industrial processes, deforestation, and certain agricultural practices also contribute to CO_2 emissions. While natural sources, like volcanic activity and respiration, release CO_2, human activities have significantly increased its concentration in the atmosphere. Carbon dioxide plays a crucial role in the Earth's climate system by contributing to the greenhouse effect. The greenhouse effect allows sunlight to enter the atmosphere and reach the Earth's surface. As the Earth absorbs this sunlight, it emits infrared radiation back into space. Greenhouse gases, including CO_2, trap some of this infrared radiation, preventing it from escaping and thus warming the planet. However, the excessive release of CO_2 from human activities has intensified the greenhouse effect, leading to global warming and climate change. The increased concentration of CO_2 and other greenhouse gases in the atmosphere results in rising global temperatures, altered weather patterns, sea level rise, and other environmental impacts. The warming caused by elevated CO_2 levels contributes to the current climate crisis, with far-reaching consequences for ecosystems, biodiversity, and human societies.

2. Methane (CH_4): It is primarily emitted from agriculture (livestock digestion and rice paddies), fossil fuel production and use, landfills, and natural sources. Methane is a potent greenhouse gas, with a higher warming potential than CO_2 over a shorter timeframe. While it persists for a shorter period, its impact on climate change is significant. Controlling methane emissions is crucial for mitigating global warming.

3. Nitrous Oxide (N_2O): It is primarily emitted from agricultural activities (fertilizer use and soil management), industrial processes, and combustion of fossil fuels. Nitrous oxide is a long-lived greenhouse gas that contributes to both global warming and ozone depletion. It has a warming potential several times that of CO_2 over a century.

4. Fluorinated Gases: It is primarily emitted from industrial processes, such as the production of refrigerants, solvents, and foam-blowing agents. While fluorinated gases have a relatively low concentration in the atmosphere, they have high warming potentials. Efforts to reduce emissions of these gases are essential for climate change mitigation.

5. Water Vapor: It is primarily emitted from evaporation from oceans, lakes, and rivers. Water vapor is a natural greenhouse gas and a significant contributor to the natural greenhouse effect. While human activities do not directly emit water vapor, they influence its levels indirectly through activities that affect other greenhouse gas concentrations.

10.9 Acid Rain and Its Consequences

Acid rain, a consequence of atmospheric pollution resulting from the release of sulfur dioxide (SO_2) and nitrogen oxides (NO*x*) into the air, poses significant threats to ecosystems and various aspects of the environment. When sulfur dioxide and nitrogen oxides are emitted from industrial activities, fossil fuel combustion, and transportation, they undergo chemical reactions in the atmosphere, producing sulfuric acid and nitric acid. These acidic compounds are then deposited onto the Earth's surface through precipitation in the form of rain, snow, fog, or dust. The consequences of acid rain are far-reaching, impacting aquatic ecosystems, soil health, and vegetation. In aquatic environments, acid rain lowers the pH of lakes and rivers, leading to acidification, which can have devastating effects on fish and other aquatic organisms, disrupting their reproductive cycles and often leading to population decline. In soils, acid rain can result in the leaching of essential nutrients such as calcium and magnesium, negatively affecting plant growth and the health of terrestrial ecosystems. Additionally, the corrosive nature of acid rain can damage buildings, monuments, and infrastructure.

10.10 Ozone Depletion in the Stratosphere

Ozone depletion in the stratosphere is a critical environmental issue that stems from the release of chlorofluorocarbons (CFCs), halons, and other ozone-depleting substances (ODS) into the atmosphere. The stratospheric ozone layer plays a crucial role in shielding life on the Earth from the harmful effects of ultraviolet (UV) radiation emitted by the sun. However, the introduction of ODS, primarily from industrial and anthropogenic activities, has led to the breakdown of ozone molecules in the stratosphere. The most significant culprits are CFCs, which, once released into the atmosphere, rise to the stratosphere where they are broken down by UV radiation, releasing chlorine atoms. These chlorine atoms then catalytically destroy ozone molecules, creating an ozone hole. The consequences of ozone depletion are profound, with increased UV radiation reaching the Earth's surface. This elevated UV radiation poses severe threats to human health, causing skin cancers, cataracts, and immune system suppression. Moreover, it has detrimental effects on ecosystems, affecting plant growth, and phytoplankton productivity, and disrupting aquatic food chains. The international response to ozone depletion was exemplified by the Montreal Protocol, a landmark agreement that aimed to phase out the production and consumption of ODS.

10.11 International Efforts and Agreements

International efforts to address air pollution and environmental issues have been formalized through various agreements and protocols, each aiming to mitigate the impact of human activities on the global environment. Two significant agreements that have garnered international attention are the Kyoto Protocol and the Paris Agreement.

1. Kyoto Protocol (1997): The Kyoto Protocol, adopted in 1997 in Kyoto, Japan, is an international treaty under the United Nations Framework Convention on Climate Change (UNFCCC). While primarily focused on reducing greenhouse gas emissions to combat climate change, the protocol indirectly addresses air pollution by targeting key pollutants associated with industrial activities. The protocol introduced legally binding emission reduction targets for developed countries, known as Annex I parties, for the commitment period 2008–2012. These targets aimed to collectively reduce emissions of six greenhouse gases, including carbon dioxide (CO_2) and methane (CH_4).

The Clean Development Mechanism (CDM) and Joint Implementation (JI) were innovative approaches under the Kyoto Protocol that allowed developed countries to invest in emission reduction projects in developing countries to meet their reduction targets.

2. Paris Agreement (2015): The Paris Agreement, adopted in 2015, represents a landmark effort to unite nations in the fight against climate change. While building on the principles of the Kyoto Protocol, the Paris Agreement takes a more inclusive approach by engaging all countries, both developed and developing, in collective efforts to limit the rise in global temperature. The agreement aims to keep the increase in global average temperature well below 2 °C above pre-industrial levels, with an aspiration to limit it to 1.5 °C.

The Paris Agreement emphasizes nationally determined contributions (NDCs), allowing each country to set its climate action goals and regularly report on its progress. This flexibility encourages a broader range of countries to participate actively in emission reduction efforts.

Both the Kyoto Protocol and the Paris Agreement reflect the international community's commitment to addressing environmental challenges collaboratively. However, the effectiveness of these agreements relies on the sustained commitment of nations to implement and strengthen their climate and air quality policies. Continuous dialogue, technological innovation, and financial support are essential components of international efforts to combat air pollution and its broader environmental impacts.

10.12 Case Studies of Successful Pollution Control Measures in Different Countries

Several countries have implemented successful pollution control measures to address various environmental challenges. Here are case studies of notable success stories from different parts of the world:

1. China – Air Quality Improvement in Beijing: China, particularly Beijing, faced severe air pollution due to rapid industrialization and urbanization. In response, the Chinese government implemented aggressive measures, including strict emission standards, increased green spaces, and a shift toward renewable energy sources. The "Airborne Pollution Prevention and Control Action Plan" launched in 2013 aimed to reduce particulate matter (PM) and improve air quality. The adoption of cleaner technologies, the promotion of electric vehicles, and enhanced public transportation infrastructure contributed to a significant improvement in air quality in Beijing.

2. Germany – Phasing Out Coal and Advancing Renewable Energy: Germany has been a pioneer in transitioning to renewable energy and phasing out coal. The "Energiewende" policy, initiated in the early 2000s, set ambitious targets for increasing the share of renewable energy in the country's total energy consumption. As a result, Germany has significantly reduced its reliance on coal and increased the use of wind and solar power. This transition not only reduced air pollution but also contributed to Germany becoming a leader in sustainable energy practices.

3. United States – The Clean Air Act and Acid Rain Control: The Clean Air Act in the United States, first passed in 1963 and amended in subsequent years, has been instrumental in controlling air pollution. One notable success is the reduction of acid rain. In the 1980s, the U.S. implemented a cap-and-trade program to limit sulfur dioxide (SO_2) emissions, a major contributor to acid rain. This program achieved substantial reductions in SO_2 emissions, leading to a significant decrease in acid rain and its adverse effects on ecosystems.

4. Sweden – Sustainable Waste Management: Sweden has implemented innovative waste management practices to minimize environmental impact. The country has adopted a "circular economy" approach, focusing on waste-to-energy incineration with energy recovery. Only a small percentage of waste ends up in landfills. This approach not only reduces the environmental impact of waste disposal but also generates energy, contributing to Sweden's renewable energy goals.

These case studies highlight the importance of comprehensive policies, technological innovation, and international collaboration in addressing pollution. Successful strategies often involve a combination of regulatory frameworks, public awareness campaigns, and incentives for industries to adopt cleaner practices. Sharing these success stories can inspire other nations to adopt effective pollution control measures tailored to their unique environmental challenges.

10.13 Mitigation and Control Strategies

Mitigation and control strategies for air pollution are essential to safeguard human health, protect the environment, and promote sustainable development. Various measures can be implemented to reduce air pollution, focusing on both the sources of pollutants and the enhancement of air quality. Here are detailed mitigation and control strategies:

1. Implementing and enforcing stringent emission standards for industries, vehicles, and power plants is crucial. Governments can set limits on the amount of pollutants that these sources are allowed to release into the air.
2. Regular monitoring and inspection of industries to ensure compliance with emission standards.
3. Encouraging the use of clean and renewable energy sources, such as solar, wind, and hydropower, reduces dependence on fossil fuels and lowers emissions.
4. Investing in research and development of cleaner technologies, such as catalytic converters for vehicles and advanced industrial processes that minimize emissions.
5. Implementation of sustainable agricultural practices to reduce emissions from farming activities.
6. Establishing efficient waste management systems, including recycling and proper disposal of hazardous waste, to prevent the release of pollutants into the air.
7. Promoting waste-to-energy initiatives to harness energy from waste materials while minimizing air pollution.
8. Planting trees and creating green spaces can act as natural air purifiers by absorbing pollutants and releasing oxygen.
9. Raising awareness among the public about the sources and health effects of air pollution.

10. Educating communities about sustainable practices and encouraging behavioral changes, such as reducing personal vehicle use and adopting energy-efficient appliances.
11. Engaging in international agreements and collaborations to address transboundary air pollution issues.
12. Sharing best practices, technologies, and expertise to collectively combat global air pollution challenges.
13. Providing financial incentives for industries and individuals adopting clean practices or technologies.
14. Imposing penalties for non-compliance with emission standards to ensure accountability.
15. Integrating air quality management into broader environmental and sustainable development policies.
16. Coordinating efforts across government agencies to address air pollution comprehensively.

10.13.1 Role of Plants in Mitigating Air Pollution

Plants play a crucial role in mitigating air pollution through several mechanisms. They act as natural air filters, absorbing pollutants such as carbon dioxide (CO_2), sulfur dioxide (SO_2), nitrogen oxides (NOx), and particulate matter (PM) through their leaves and stems. Trees and plants also release oxygen (O_2) as a byproduct of photosynthesis, improving air quality and supporting life. Grasses, for instance, are highly effective at absorbing harmful pollutants. Species like vetiver grass and switchgrass have been shown to absorb heavy metals and other toxins from the soil and air, making them valuable for phytoremediation efforts. These grasses can trap airborne pollutants and prevent them from spreading, thus improving air quality in their vicinity. Leaves of plants play a significant role in catching dust and other particulate matter. The hairy and waxy surfaces of leaves, such as those of the London plane tree and the common ivy, can trap dust particles and prevent them from re-entering the air. This ability to capture and hold particulate matter reduces the concentration of pollutants in the atmosphere and promotes cleaner air.

Additionally, plants help reduce the urban heat island effect by providing shade and releasing water vapor through transpiration, which cools the surrounding air. This cooling effect decreases the need for air conditioning, reducing energy consumption and the emission of greenhouse gases from power plants. Urban green spaces, such as parks and green roofs, significantly improve air quality in cities by trapping dust, soot, and other airborne particles. The presence of greenery also encourages biodiversity, which can lead to healthier ecosystems and further enhance the air-purifying capabilities of plants.

These strategies, when implemented comprehensively and collaboratively, can significantly mitigate air pollution, leading to improved air quality and enhanced overall well-being for both current and future generations.

Chapter 11

Water Pollution

"A call for action to protect our sources of life"

11.1 Introduction

Water is the essential element that makes life on the earth possible. Without water, there would be no life. Although 71% of the earth's surface is covered by water, only a tiny fraction of this water is available to us as fresh water.

About 97% of the water on the earth is found in oceans and is too salty for drinking or irrigation. The remaining 3% is fresh water. Of this, 2.997% is locked in ice caps or glaciers. Thus only 0.003% of the earth's total volume of water is easily available to us as soil moisture, groundwater, water vapor, and water in lakes, streams, rivers, and wetlands. Water in streams, rivers, lakes, wetlands, and artificial reservoirs is called surface water. Water that percolates into the ground and fills the pores in soil and rock is called groundwater. Porous water-saturated layers of sand, gravel, or bedrock through which groundwater flows are called aquifers. Most aquifers are replenished naturally by rainfall that percolates downward through the soil and rock. This process is called natural recharge. If the withdrawal rate of an aquifer exceeds its natural recharge rate, the water table is lowered. Any pollutant discharged onto the land above

Environmental Toxicology: Concepts and Principles
Mohd Salim Reshi, Darakhshan Javaid, Syed Sanober Qadri, and Neelu Sinha

ISBN 978-981-5129-68-7 (Hardcover), 978-1-003-63707-3 (eBook)
www.jennystanford.com

is also pulled into the aquifer and pollutes the groundwater resulting in polluted water in the nearby wells.

11.2 Definition

When the quality or composition of water changes directly or indirectly because of man's activities such that it becomes unfit for any purpose it is said to be polluted or any physical, biological, or chemical change in water quality that adversely affects living organisms or makes water unsuitable for desired use is called water pollution. It could be due to natural or human activities. Natural sources of water pollution are soil erosion, leaking of minerals from rocks, and decaying organic matter, while human-made sources include domestic, agricultural, and industrial activities. Many water sources have become a dumping ground for wastes.

11.3 Sources of Water Pollution

There are two sources of water pollution based on the origin of pollutants: (i) point sources and (ii) non-point sources.

1. Point Sources: When a source of pollution can be readily identified because it has a definite source and place where it enters the water. It comes from point sources like municipal and industrial discharge pipes. Oil wells near water bodies, factories, power plants, underground coal mines, etc. are point sources of water pollution.

2. Non-point Sources: When a source of pollution cannot be readily identified, such as agricultural runoff, acid rain, etc., they are said to be non-point sources of pollution. Those sources which do not have any specific location for discharging pollutants, in the water body are known as non-point sources of water pollution. Run-offs from agricultural fields, lawns, gardens, construction sites, roads, and streets are some non-point sources of water pollution.

11.4 Types of Water Pollutants

There are several classes of common water pollutants. These are disease causing agents (pathogens), which include bacteria, viruses, protozoa, and parasitic worms that enter water from domestic sewage and untreated human and animal wastes.

Another category of water pollutants is oxygen-depleting wastes. These organic wastes can be decomposed by aerobic (oxygen-requiring) bacteria. The amount of oxygen required to break down a certain amount of organic matter is called the biological oxygen demand (BOD). The amount of BOD in the water is an indicator of the level of pollution. If too much organic matter is added to the water all the available oxygen is used up. This causes fish and other forms of oxygen-dependent aquatic life to die. Thus, anaerobic bacteria (those that do not require oxygen) begin to break down the wastes. Their anaerobic respiration produces chemicals that have a foul odor and an unpleasant taste that is harmful to human health.

A third class of pollutants is inorganic plant nutrients. These are water-soluble nitrates and phosphates that cause excessive growth of algae and other aquatic plants. The excessive growth of algae and aquatic plants due to added nutrients is called eutrophication. They may interfere with the use of the water by clogging water intake pipes, changing the taste and odor of water, and causing a buildup of organic matter. As the organic matter decays, oxygen levels decrease and fish and other aquatic species die.

A fourth class of water pollutants is water-soluble inorganic chemicals which are acids, salts, and compounds of toxic metals such as mercury and lead. High levels of these chemicals can make the water unfit to drink and harm.

Another cause of water pollution is a variety of organic chemicals, which include oil, gasoline, plastics, pesticides, cleaning solvents, detergents, and many other chemicals.

Sediment of suspended matter is another class of water pollutants. These are insoluble particles of soil and other solids that become suspended in water. Excessive sediments that settle down destroy the feeding and spawning grounds of fish, clog and fill lakes, artificial reservoirs, etc.

Water-soluble radioactive isotopes are yet another source of water pollution. These can be concentrated in various tissues and organs as they pass through food chains and webs.

Hot Water: Let out by power plants and industries that use large volumes of water to cool the plant resulting in a rise in the temperature of the local water bodies. This heated water, which is at least 15 °C higher than normal is discharged back into the water body. The warm water not only decreases the solubility of oxygen but changes the breeding cycles of various aquatic organisms.

Pesticides: Pesticides protect crops against diseases and plagues and normally also increase yield and the quality of the product. Present intensive agriculture, sometimes even based on monocultures, poses a major risk of diseases and plagues. There is an increasing imbalance in the formerly ecologically sound agro-ecosystems. In Germany, Norway, Sweden, the United States, and some other countries, pesticide residues have been found in samples of groundwater and surface water in quantities exceeding drinking-water standards and/or ecologically based water-quality criteria and objectives.

11.5 Indicators of Water Pollution

Indicators of water pollution are critical for assessing the health and safety of water bodies. These indicators can be broadly categorized into physical, chemical, and biological parameters.

1. **Physical Indicators:**
   - **Turbidity:** Refers to the cloudiness or haziness of water caused by suspended particles. High turbidity can reduce light penetration, affecting aquatic life and indicating the presence of pollutants.
   - **Color:** Discoloration of water can indicate contamination from organic matter, chemicals, or industrial discharges.
   - **Temperature:** Elevated water temperatures can reduce dissolved oxygen levels and harm aquatic organisms. It can be a sign of thermal pollution from industrial processes.
2. **Chemical Indicators:**
   - **pH Levels:** The acidity or alkalinity of water. Extreme pH levels can harm aquatic life and indicate chemical pollution.
   - **Dissolved Oxygen (DO):** The amount of oxygen dissolved in water. Low DO levels can be a sign of organic pollution, which consumes oxygen during decomposition.

- **Nutrients:** High levels of nitrates and phosphates can lead to eutrophication, causing algal blooms and oxygen depletion.
- **Heavy Metals:** The presence of metals like lead, mercury, and cadmium can be toxic to aquatic life and humans.
- **Chemical Oxygen Demand (COD) and Biological Oxygen Demand (BOD):** High COD and BOD values indicate high levels of organic pollutants, which consume oxygen during decomposition.

3. **Biological Indicators:**
   a. **Coliform Bacteria:** The presence of coliforms, particularly *E. coli*, indicates fecal contamination and the potential presence of pathogens.
   b. **Macroinvertebrates:** Certain species of insects, crustaceans, and worms are sensitive to pollution. Their presence or absence can indicate water quality.
   c. **Algal Blooms:** Excessive growth of algae, particularly harmful algal blooms (HABs), can indicate nutrient pollution and lead to toxins harmful to aquatic life and humans.

Monitoring these indicators helps identify the sources and extent of water pollution, allowing for effective management and remediation efforts to protect aquatic ecosystems and human health.

11.6 International Standards for Drinking Water

1. Fluoride: For drinking purposes, water should be tested for fluoride ion concentration. Its deficiency in drinking water is harmful to man and causes diseases such as tooth decay, etc. Soluble fluoride is often added to drinking water to bring its concentration up to 1 ppm or 1 mg/l. However, F^- ion concentration above 2 ppm causes brown mottling of teeth. At the same time, excess fluoride (over 10 ppm) causes harmful effects on bones and teeth.

2. Lead: Drinking water gets contaminated with lead when lead pipes are used for the transportation of water. The prescribed upper limit concentration of lead in drinking water is about 50 ppb. Lead can damage the kidney, liver, reproductive system, etc.

3. Sulfate: Excessive sulfate (>500 ppm) in drinking water causes the laxative effect, otherwise at moderate levels it is harmless.

The maximum limit of nitrate in drinking water is 50 ppm. Excess nitrate in drinking water can cause diseases such as methemoglobinemia ('blue baby' syndrome).

11.7 Impacts of Water Pollution

1. Human Health Risks: Water pollution poses substantial risks to human health, primarily through the contamination of drinking water sources. Inadequately treated or polluted water can harbor a myriad of pathogens, including bacteria, viruses, and parasites, leading to waterborne diseases. Cholera, dysentery, and gastroenteritis are just a few examples of illnesses that can result from consuming contaminated water. The consequences are particularly severe in developing regions with limited access to clean water and sanitation facilities. Prolonged exposure to chemical pollutants in water, such as heavy metals and industrial chemicals, can also lead to chronic health issues, including organ damage and neurological disorders. Safeguarding water quality is paramount for preventing the spread of waterborne diseases and protecting the well-being of communities worldwide.

2. Ecological Damage: The ecological implications of water pollution are extensive and impact diverse aquatic ecosystems. Contaminants like chemical pollutants, nutrients, and microbial agents disrupt the delicate balance of marine and freshwater environments. In the case of nutrient pollution, the excessive input of nitrogen and phosphorus fuels algal blooms, leading to oxygen depletion and the creation of "dead zones" where marine life cannot thrive. Chemical pollutants, including pesticides and heavy metals, accumulate in the tissues of aquatic organisms, causing physiological imbalances and threatening the biodiversity of rivers, lakes, and oceans. The disruption of food chains and habitats further leads to ecological imbalances, emphasizing the urgent need for comprehensive measures to mitigate the impact of water pollution on ecosystems.

3. Economic Consequences: Water pollution takes a significant toll on economies, affecting various sectors and leading to substantial economic losses. Industries reliant on clean water sources, such as fisheries and tourism, face direct impacts as contaminated waters degrade the quality and quantity of available resources. Fishery yields decline due to the adverse effects of pollutants on aquatic life, risking livelihoods, and food security. Tourism suffers as polluted water bodies lose appeal, with potential visitors deterred by environmental degradation. Moreover, the costs associated with cleaning up polluted water sources and treating water for consumption place a considerable financial burden on governments and local communities. Preventing water pollution is not merely an environmental imperative but also an economic necessity for the sustainable development of communities and nations.

4. Drinking Water Scarcity: Water pollution exacerbates the global challenge of drinking water scarcity, as contamination renders vast water sources unfit for consumption. In many regions, communities rely on surface water from rivers and lakes, which are susceptible to pollution from agricultural runoff, industrial discharges, and untreated sewage. As the population grows and industrialization advances, the demand for clean water intensifies, making pollution prevention and water management critical.

11.8 Case Studies and Notable Incidents of Water Pollution and Its Consequences

1. Minamata Mercury Poisoning: One of the most infamous cases of water pollution is the Minamata mercury poisoning incident in Japan during the mid-20th century. The Chisso Corporation, a chemical company, discharged methylmercury into Minamata Bay as part of its industrial waste. The mercury bioaccumulated in fish and shellfish, which were staple foods for the local population. Tragically, consumption of contaminated seafood led to severe neurological disorders, birth defects, and even death among residents. The incident highlighted the devastating human health impacts of industrial water pollution and catalyzed global awareness about the dangers of mercury contamination.

2. Love Canal Contamination: The Love Canal incident in the United States, particularly in Niagara Falls, New York, is a landmark case illustrating the dangers of improper disposal of chemical waste. In the 1940s and 1950s, Hooker Chemical Company disposed-off toxic chemical waste, including carcinogens like benzene and dioxin, in an abandoned canal. In the 1970s, a residential neighborhood was built on the site, and residents began experiencing health issues. The area was declared a federal emergency as investigations revealed widespread contamination of soil and groundwater. The Love Canal disaster prompted the creation of the Superfund program in the U.S., aimed at cleaning up hazardous waste sites and preventing similar incidents.

3. The Citarum River, Indonesia: The Citarum River, considered one of the most polluted rivers globally, exemplifies the impact of industrial and domestic pollution on water quality. Situated in West Java, Indonesia, the river receives discharges from hundreds of textile factories, tanneries, and households. The water is contaminated with heavy metals, toxic chemicals, and untreated sewage. The consequences are terrible for both the environment and human health. Aquatic life struggles to survive, and local communities relying on the river for drinking water face severe health risks. Efforts to clean up the Citarum River highlight the challenges of addressing water pollution in densely populated and industrialized regions.

4. Flint Water Crisis: The Flint water crisis in Michigan, USA, is a stark example of how inadequate water treatment and management can lead to public health emergencies. To save money, the city switched its water source to the Flint River in 2014, but the water was not properly treated for corrosion control. This resulted in lead leaching from aging pipes into the drinking water, exposing thousands of residents, including children, to elevated lead levels. The crisis emphasized the importance of maintaining water infrastructure and the potential health consequences when proper precautions are not taken.

These case studies underscore the urgency of addressing water pollution through effective regulation, responsible industrial practices, and community engagement. They serve as cautionary tales, emphasizing the need for proactive measures to prevent and mitigate the far-reaching consequences of water contamination.

11.9 Control Measures for Preventing Water Pollution

- Setting up effluent treatment plans to treat waste
- Water recycling
- Treatment of industrial wastes before discharge
- Chemical pesticides and herbicides are very harmful to the water and the environment. There must be safer replacements for these products
- Cut back on bleaches and detergents
- Educate people about preventing water pollution and the consequences of water pollution
- Strict enforcement of the Water Pollution Control Act
- Continuous monitoring of water pollution at different places
- Developing economical methods of water treatment

11.10 Treatment of Sewage

The sewage can be treated by a modern technique involving three steps—primary treatment, secondary treatment, and tertiary treatment.

In primary treatment, sewage is passed through a grinding mechanism. This is then passed through several settling chambers and lime is added to neutralize it. The neutralized sewage still contains many pathogenic and non-pathogenic organisms and enough organic matter.

In the secondary treatment, these neutralized effluents are passed through a reactor called UASB (upflow anaerobic sludge blanket). In this reactor, the anaerobic bacteria degrade the biodegradable material into neutralized effluents. In this process, the foul odor and methane are released and the sewage is converted into clean water. This water is sent to aeration tanks where air and bacteria are added to it. This process is called biological or secondary treatment.

The water obtained because of secondary treatment is still unfit for drinking and needs further purification. This is done by tertiary treatment. In this treatment, which is a disinfecting process, final traces of disinfecting bacteria and any dissolved absorption methods are employed to obtain clean water.

11.11 Ongoing Challenges in Addressing Water Pollution

1. Lack of Comprehensive Regulatory Frameworks: Inconsistencies in environmental regulations and enforcement capacities globally hinder the effective control of water pollution. Variations in standards across regions allow some areas to engage in environmentally harmful practices without adequate repercussions.

2. Rapid Urbanization and Industrial Growth: The swift pace of urbanization and industrial expansion contributes to increased water pollution. Urban runoff, inadequate waste management, and intensified industrial activities result in the discharge of pollutants into water bodies, putting a strain on natural purification processes.

3. Agricultural Practices and Runoff: The widespread use of fertilizers and pesticides in agriculture leads to nutrient runoff, causing eutrophication and contaminating water bodies. Agricultural practices, often driven by the need for higher yields, contribute significantly to water pollution.

4. Limited Access to Clean Water: Access to clean and safe drinking water remains a challenge in many parts of the world. As populations grow, particularly in developing regions, the demand for water increases, leading to over-extraction, contamination, and inadequate sanitation infrastructure.

5. Emerging Contaminants: Identifying and regulating emerging contaminants, such as pharmaceuticals and microplastics, present ongoing challenges. The potential long-term effects of these pollutants are not fully understood, complicating efforts to control and mitigate their impact.

11.12 Future Prospects for Water Quality Improvement

- Ongoing advancements in water treatment technologies hold promise to improve water quality. Innovations in filtration, desalination, and wastewater treatment technologies can contribute to minimizing and removing pollutants from water sources.

- Strengthening international cooperation and governance frameworks is essential for addressing water pollution collectively. Harmonizing environmental standards, sharing best practices, and establishing effective monitoring and enforcement mechanisms can create a more coordinated global effort.
- Empowering communities through education and awareness programs is crucial for long-term success. Informed communities are more likely to adopt sustainable practices, demand better governance, and actively participate in pollution prevention efforts.
- Implementing nature-based solutions, such as wetland restoration, afforestation, and green infrastructure, can play a pivotal role in mitigating water pollution. These approaches leverage natural processes to enhance water quality and provide additional ecological benefits.
- Transitioning toward a circular economy, where resources are used more efficiently, can contribute to reducing pollution. This involves promoting recycling, minimizing waste, and designing products with a focus on their entire life cycle, thus lessening the environmental impact.
- Governments and policymakers need to continually adapt and innovate their strategies to address evolving pollution challenges. This includes updating regulations, incentivizing sustainable practices, and investing in research to understand and address emerging threats to water quality.
- Collaborative efforts between governments, private industries, and non-governmental organizations (NGOs) can provide the necessary resources and expertise to tackle water pollution. Public-private partnerships can drive innovation and create sustainable solutions with a focus on shared responsibility.

11.13 Role of Plants in Mitigating Water Pollution

Plants play a crucial role in mitigating water pollution through various mechanisms that enhance water quality and promote ecosystem health. Here are some of the key ways plants contribute to water pollution control:

1. **Phytoremediation:**
   - **Absorption and Uptake:** Certain plants, known as hyperaccumulators, can absorb and concentrate heavy metals and other pollutants from contaminated water through their roots and shoots. This process reduces the levels of harmful substances in the water.
   - **Rhizofiltration:** Aquatic plants and their root systems can filter out contaminants such as heavy metals and excess nutrients from the water, effectively cleaning it as it passes through the root zone.
2. **Reduction of Nutrient Pollution:**
   - **Nutrient Uptake:** Plants, especially aquatic plants like water hyacinth, duckweed, and cattails, can absorb excess nutrients such as nitrogen and phosphorus from the water. This helps prevent nutrient pollution, which can lead to harmful algal blooms and oxygen depletion.
   - **Buffer Strips**: Riparian buffer strips, consisting of grasses, shrubs, and trees planted along water bodies, can trap and utilize nutrients from agricultural runoff before they enter the water, reducing the impact of nutrient pollution.
3. **Erosion Control and Sediment Trapping:**
   - **Root Systems:** The extensive root systems of plants help stabilize soil and reduce erosion. By preventing soil erosion, plants help keep sediments out of water bodies, which can carry pollutants and increase turbidity.
   - **Vegetated Buffers**: Plants along riverbanks and shorelines act as barriers that trap sediments and pollutants from surface runoff, preventing them from reaching water bodies.
4. **Enhanced Biological Activity:**
   - **Habitat for Microorganisms:** Plant roots provide a habitat for beneficial microorganisms that can degrade organic pollutants, such as hydrocarbons and pesticides, through microbial processes.
   - **Oxygen Production:** Aquatic plants produce oxygen during photosynthesis, which can improve dissolved oxygen levels in the water, supporting aquatic life and enhancing the breakdown of organic pollutants.

5. **Natural Filtration Systems:**
   - **Constructed Wetlands:** Engineered systems that mimic natural wetlands use a combination of plants, soil, and microorganisms to treat wastewater. These systems can effectively remove a wide range of pollutants, including heavy metals, pathogens, and nutrients.
   - **Green Infrastructure:** Urban green infrastructure, such as rain gardens and green roofs, utilizes plants to filter and absorb stormwater runoff, reducing the load of pollutants entering water bodies.

By leveraging these natural processes, plants play an essential role in maintaining and improving water quality, providing a sustainable and cost-effective solution for water pollution management.

In conclusion, while the challenges in addressing water pollution persist, there are optimistic prospects for improvement. Technological advancements, strengthened governance, community engagement, and a shift toward sustainable practices offer hope for a future where water quality is safeguarded for the benefit of both ecosystems and human populations.

Chapter 12

Ecotoxicology

"Nature's balance shattered by toxins"

12.1 Introduction

The word "ecotoxicology" is a combination of two words "ecology" and "toxicology". Ecotoxicology is concerned with the study of potentially harmful chemicals (toxins) and their effects on the ecosystem. Ecotoxicology needs a proper understanding of ecological principles and theory, which will help assess how chemicals can affect individuals, populations, communities, and ecosystems.

Ecology serves as the foundation for ecotoxicology, offering a framework for developing ecotoxicological theory. Positioned within a broader scope of environmental toxicology, ecotoxicology focuses primarily on understanding the causes and consequences of toxic substance exposure in ecosystems, particularly at higher organizational levels. Environmental factors inherent to natural settings play a crucial role in regulating or influencing the life cycles of organisms, genetic variability, and impacts of toxic substances. Consequently, adopting an ecological perspective in toxicology enhances the precision of assessing toxic substance levels (chemical concentrations) in the environment. This refined

Environmental Toxicology: Concepts and Principles
Mohd Salim Reshi, Darakhshan Javaid, Syed Sanober Qadri, and Neelu Sinha

ISBN 978-981-5129-68-7 (Hardcover), 978-1-003-63707-3 (eBook)
www.jennystanford.com

ecotoxicological knowledge becomes valuable for evaluating the severity of pollution and determining, with reasonable accuracy, the maximum permissible concentrations for substance emissions into the environment.

Ecotoxicology grapples with intricate systems, making it impractical to comprehend all the differences of every ecological component and process implicated in ecotoxicological issues. However, it is feasible to employ various theories from contemporary ecology to scrutinize environmental problems, focusing on, at the very least, the most crucial components and processes. Hoffman et al. (1995) have outlined ecotoxicological methods and procedures. This approach hinges on the relationship between an individual organism and environmental factors, considering biological aspects such as uptake and excretion rates, biological concentration factors, and ecological magnification factors.

The aquatic and terrestrial ecosystems have different parameters to regulate the life cycle of organisms and their toxic effects. Therefore, the ecotoxicology for both aquatic and terrestrial ecosystems should be studied separately. This chapter addresses ecology, aquatic, and terrestrial ecotoxicology.

12.2 Ecosystem

The biotic community along with its habitat is known as an ecosystem (Tansley, 1935). A community is an aggregate of organisms, which form a distinct ecological unit. Thus, the ecosystem may be defined as a specific biological community and its physical and chemical environment interacting with each other. In simple terms, an ecosystem includes and sustains life. Generally, an ecosystem includes the life of at least an autotroph, a consumer, a decomposer, and life-sustaining elements like air, water, soil, energy, and other essential elements. Habitat is a specific set of physical and chemical conditions (e.g., space, substratum, climate) that surround a single species, a group of species, or a large community (Clements and Shelford, 1939). The ultimate division of the biosphere is the microhabitat and greater than the microhabitat is called macrohabitat. Microhabitats are mostly intimately local and immediate conditions surrounding an organism, e.g., the borrow of a rodent. Thus, an ecological community is a set of interrelated species that occur at the same place. Since plants, animals, bacteria, fungi, etc., occur together in the same habitat and are interrelated, they cannot survive independently.

The composition and character of the biotic community is the strong indicator of the type of environment that is present. The biotic community may be grouped into two parts based on their sustaining ability: (i) major community and (ii) minor community.

Major communities are those which along with their habitats form more or less complete and self-sustaining ecosystems (sunlight is excluded). The minor communities are secondary aggregations within major communities and are not completely independent ecosystems. The biological ecosystems have three important characteristics: (i) structure, (ii) processes, and (iii) succession.

12.3 Ecosystem Structure and Community Dynamics

The ecosystem may be natural or man-made ecosystems. The natural ecosystems may be further categorized as: (i) terrestrial and (ii) aquatic ecosystems. These ecosystems are further subdivided as shown in the flow chart (Fig. 12.1).

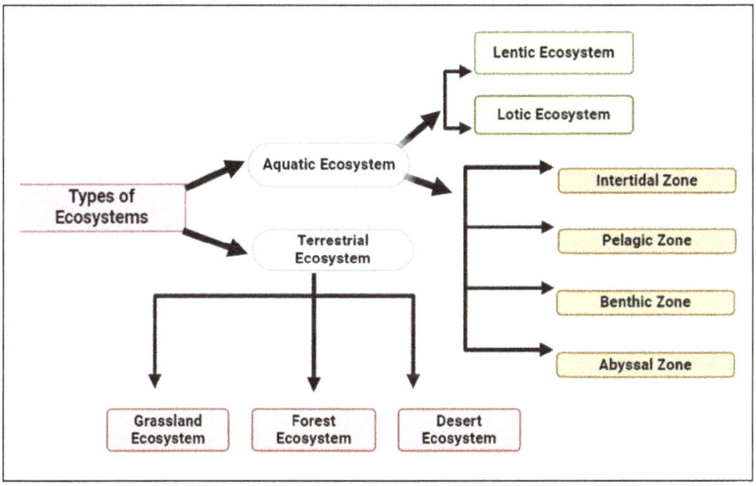

Figure 12.1 The flow diagram of the ecosystem.

Each ecosystem has two important components: (i) biotic component (living organisms) and (ii) abiotic component (non-living things). The biotic component is further subdivided into three important units such as producers, consumers, and decomposers. Similarly, the abiotic component is further subdivided into physical and chemical components shown in Fig. 12.2.

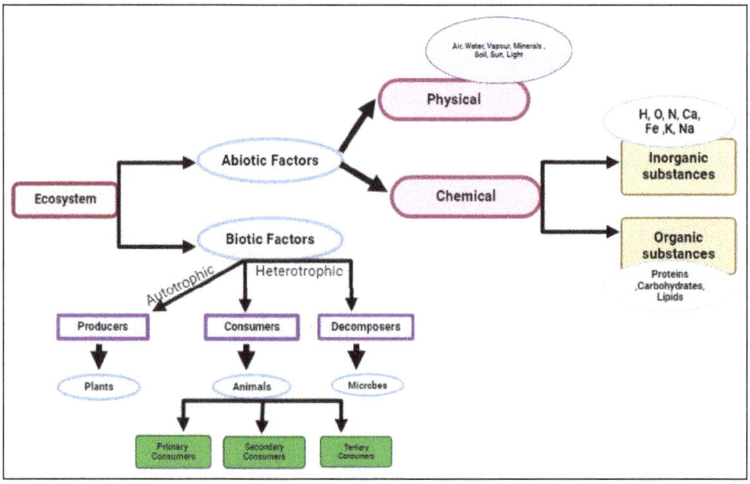

Figure 12.2 Flow diagram of components of the ecosystem.

12.3.1 Terrestrial Ecosystem

The terrestrial ecosystem refers to an ecosystem other than the ecosystem of the water body, i.e., the terrestrial ecosystem consists of three important biotic provinces or ecosystems: (i) the grassland ecosystem, (ii) the forest ecosystem, and (iii) the desert ecosystem.

A biotic province is a region inhabited by a characteristic set of species or families bound by barriers that prevent the spread of distinctive kinds of life to other regions and the immigration of foreign species.

12.3.1.1 Grassland ecosystem

Grasslands occur in temperate regions, which have moderately dry climates. The grassland vegetation is completely renewed each year. Grasses may be divided into three categories based on their height:

(i) tall grasses (1.5–3 m), (ii) mid grasses (0.5–1.5 m), and (iii) short grasses (<0.5 m).

The grassland community typically encompasses subterranean, surface, and herbaceous strata of vegetation. In temperate grasslands, the prevalent species include grasses and other flowering plants, many of which are perennial and characterized by well-established root systems. The nomenclature for grasslands varies across regions, and they are known by different names in different areas (Table 12.1).

Table 12.1 Region and grassland

Region	Grassland
North America	Prairies
Eurasia	Steppes
South America	Pampas
South Africa	Veldt

Because of the dry habitat of grasslands, insects are abundant. Most populations are restricted to soil, litter, and shrubs as grasslands have few trees because of inadequate rainfall. These insects include spiders, ants, leaf hoppers, millipedes, earthworms, etc. Birds are not numerous in grassland. Sparrows, marsh hawks, owls, etc. are generally found in the grassland ecosystem. Secondary consumers like foxes, wolves, jackals, snakes, and hawks are also found in the grassland ecosystem. Grazing animals (primary consumers) are also found in abundance such as cows, buffaloes, deer, sheep, goats, horses, zebra, and kangaroos.

12.3.1.2 Forest ecosystem

Climatic factors, particularly temperature and precipitation, play a pivotal role in determining the distribution of biomes within the forest ecosystem. Forests are characterized by the presence of trees growing in close proximity, effectively dominating the entire ground surface. Forest vegetation can be broadly categorized into two types: closed canopy and open woodland forests. In a closed canopy forest, the leaves and twigs of neighboring trees make contact with each other, creating a dense cover. Conversely, in an open woodland, only certain leaves and twigs of adjacent trees overlap, resulting in a more dispersed and less densely packed canopy (Fig. 12.3).

Figure 12.3 Closed canopy and open woodland (Source: Internet).

According to the climatic conditions, the composition of the forest may vary. Approximately one-third of the land is covered by forest. Forest in terms of ecology includes a complex organism comprising distinct biological units (plants and animals) called forest communities. In warm and moist climates, the forests are mostly broad-leaved deciduous; in cold climates, they are mostly needle-leaved evergreen; and in continuously warm and moist climates, the forests are broad-leaved evergreen. Based on the climatic conditions, the forest ecosystem may be broadly grouped into the following forest ecosystems:

(i) Deciduous Forest
(ii) Coniferous Forest
(iii) Woodland
(iv) Tundra

1. Deciduous Forest: Temperate deciduous forests thrive in regions characterized by hot and wet summers. These forests are distinguished by deciduous trees that shed their leaves in the autumn, remain bare throughout winter, and sprout new foliage in the spring. The climax community of the deciduous forest biome is dominated by broad-leaved trees. A key distinction of deciduous trees from others lies in the rapid and relatively complete decomposition of their broad leaves, contributing to the formation of a nutrient-rich humus that gradually blends with the mineral soil. This biome is well-developed in Eastern North America, Western Europe, and Eastern Asia, where the trees typically create relatively dense forests with a closed canopy.

The deciduous trees shed their leaves during autumn which consequently affects the seasonal change in forest microclimates, to which animal life must respond. In these forests, large mammals are less but birds and insects are abundant. The dominant animals are small mammals including those living in trees such as monkeys, squirrels, mice, etc.

2. Coniferous Forests: Coniferous forests are predominantly found in the northern hemisphere and are characterized by a continuous, often dense, growth of needle or scale-leaved evergreen trees. The needle-like evergreen leaves, equipped with a thick waxy coating, serve to mitigate excessive water evaporation during summer and dry periods and are adapted to withstand harsh winters. These forests thrive in areas with low precipitation, reduced moisture content, and a short summer growing season. Fire plays a significant and recurring role in coniferous forests, as the waxy dead, dry needles that adhere to the trees contribute to the fuel for destructive crown fires.

The dominant conifers are pine, hemlock, spruce, cedar, larches, poplars, firs, and birches. The dominant animals are shrews, flying squirrels, wolf, fox, bear, lynx, deer, etc. The basic difference between coniferous and deciduous forests is the needle-waxy leaves of conifers decompose slowly and form a somewhat acid humus sharply defined from the underlying mineral soil.

3. Woodland: Woodland is an open stand of trees where the leaves and twigs of some trees only overlap. In woodlands, trees are short in height. They may form a closed canopy but they are scattered so the forest floor is continuously covered with grasses. The trees vary widely in leaf structure, but nearly all species are adapted to the harsh summer. These vegetation covers are evergreen and they are capable of doing photosynthesis in poor conditions also (in dry summer).

The dominant vegetations are pinyon pines and evergreen oaks. The trees are scattered, allowing considerable sunlight to reach the ground. Fire is a common disturbance in woodlands.

The animals like mule deer, mountain lions, rock squirrels, desert rats, foxes, and wolves are generally found in woodlands. Certain bird species like woodpeckers, band-tailed pigeons, flycatchers, and bluebirds are found in woodland areas.

4. Tundra: Tundras are found at far northern and southern latitudes. Tundra comprises the communities in the zone of perpetual snow and ice. The climatic conditions of the tundra are too harsh for trees. Thus, tundras are treeless plains that occur under harsh climates of low rainfall, high snowfall, and low average temperatures.

The tundra is characterized by a landscape dominated by short grass, sedges, rushes, lichens, mosses, ericaceous plants, and some flowering species. Due to challenging conditions for seed germination, most tundra plants are perennials. There are two primary types of tundra: (i) arctic tundra, which occurs at high latitudes, and (ii) alpine tundra, which is found at high elevations.

The arctic tundra environment is exceptionally harsh, and its dominant fauna includes species such as the Arctic hare, Collared lemming, Gray wolf, Arctic fox, Polar bear, and Ermine. Avian species in the arctic tundra encompass the snowy owl, willow, and rock ptarmigans. This unique ecosystem showcases a range of adaptations by flora and fauna to endure the challenging conditions of the tundra biome.

The largest alpine areas of the world lie on the Tibet plateau and in the adjacent Himalayan mountains of Asia. The thin mountain air permits intense solar radiation thus many alpine plants have deep pigmentation that shields their inner cells. Mountain goat, mountain sheep, and white-tailed ptarmigan are the dominant animals of this region.

12.3.1.3 Desert ecosystem

Deserts, ecologically speaking, encompass arid landscapes with substantial vegetation in the form of bushes, shrubs, and trees specially adapted to endure hot and dry climates. Typically, deserts occur in regions where annual rainfall is less than 50 cm. They are distributed in belts at similar latitudes both north and south of the equator, covering approximately one-fifth of the earth's surface. Prominent deserts are found in the Southwestern United States, the Sahara, Southern Africa, Arabia, Central Asia, and Australia.

Deserts are broadly classified into two types: hot deserts and cold deserts. Hot deserts are characterized by low moisture levels, minimal precipitation, and high evaporation rates, with daytime temperatures often exceeding 40 °C. Examples include the Sahara in Africa and the Thar in India. On the other hand, cold deserts exhibit lower evaporation rates, higher moisture levels, and annual temperatures around 10 °C or less. These deserts are situated in high latitudes, such as the deserts of Western Asia, and areas like the Nevadas may be categorized as cold deserts. The distinction between hot and cold deserts is based on temperature, moisture levels, and evaporation rates.

Animals in the desert have both structural and behavioral adaptations to meet hot and dry climates in terms of water deficiency and high heat. Many desert animals escape the daytime and hide themselves in burrows or rocky shelters to avoid high temperatures. The dominant animals are ground squirrels, jackrabbits, kangaroo rats, pocket mice, foxes, grasshoppers, etc. Among birds, the dominant are hawk, hummingbird, woodpecker, sparrow, and flycatcher.

Soils often have abundant nutrients but little or no organic matter and need only water to become very productive. Leaves of plants are reduced to scales or spines to reduce the transpiration of water. The dominant plant species in hot deserts are cacti, creosote bush, yucca, and other xerophyte species. In the cold deserts, perennial plants grow to survive under very harsh cold conditions. In the desert areas, the net primary production is very low.

12.3.2 Aquatic Ecosystem

Water is essential for the survival of life in every biological community and serves as an important habitat for many organisms. The aquatic ecosystems may be classified as:

(i) Freshwater ecosystems
(ii) Saline ecosystems

12.3.2.1 Freshwater ecosystems

Freshwater ecosystems include the standing waters of ponds, wells, groundwater, and lakes, and the flowing waters of rivers and streams. The aquatic habitat offers a variety of unique conditions to which organisms must be adopted before they can occur. The habitat conditions for standing water and flowing water are different. The aquatic habitats may be classified as:

(i) Lotic Habitat: Lotic habitat is essentially the habitat of the continuous and rapid flow of water bodies, for example, streams, rivers, and springs.

(ii) **Lentic Habitat**: In lentic habitat, water is essentially a standing quiescent body, for example, ponds, lakes, and wells.

Aquatic organisms need carbon dioxide, water, and sunlight for photosynthesis; oxygen for respiration; and food and mineral nutrients for energy, growth, and maintenance. The availability of these above necessities depends on:

(i) Solubility in water (oxygen, minerals, etc.)
(ii) Depth of water body
(iii) Penetration of light and temperature
(iv) Flow rate of a water body
(v) Bottom characteristics

12.3.2.1.1 *Lotic ecosystem*

Rainwater falls on an uneven surface, it collects in depressions. As the water overflows, it will become a **stream**. When the stream is less than 3 m wide, they are called **creeks** or **brooks**. Rivers are streams of 3 m or wide. The river has three stages, namely, youth, mature, and old age. At a young age, the rivers are narrow and steep-sided. The water current is very high. At the mature age, it becomes wider and has gentle slopes and the current is slower. In old age, the river system has reached base level and the flow of water is sluggish.

Habitats

Streams harbor crucial habitats, each playing a distinct role in the aquatic ecosystem. These habitats include falls, rapids, riffles, sand bottom pools, and mud bottom pools. The characteristics of the bottom habitats are largely influenced by the volume and water current of the stream flow.

The stream bottom, featuring cobbles and boulders too heavy to be moved by the water current, is known as the **rapid habitat.** In areas where the stream bottom is turbulent, it is referred to as the **riffle habitat. Sand bottom pools** form when the water current is moderate, while **mud bottom pools** emerge when the water current at the bottom is negligible.

Mud bottom pools are particularly noteworthy for their high primary productivity, attributed to the significant presence of organic matter. This organic-rich environment fosters a thriving ecosystem, contributing to the overall biodiversity and ecological dynamics within the stream.

Stream biocies

Generally, plants are not abundant in the stream biocies, but to some extent, they are found at the upper surfaces of rocks with branched filamentous algae and a few species of water mosses (Fontinalaceae) may occur.

The most characteristic and abundant animals in the aquatic ecosystem are caddisfly larvae, mayfly naiads, stonefly naiads, crayfish, trout, snails, sponges, clams, etc. Planktons are abundantly found in sluggish water but are very rare in running water.

The nature of a stream is a pivotal factor in determining the types and abundance of species in the lotic aquatic ecosystem. Organisms in streams must adapt to the effects of water currents through clinging mechanisms, avoidance behaviors, or vigorous swimming. The segregation of stream animals among riffles, sand, and mud bottom pools can be a response to the specific characteristics of the stream bottom.

Animals inhabiting mud bottom pools are typically less equipped to deal with strong currents. They must adapt to lower oxygen concentrations in the water, as oxygen levels vary across different stream environments. Oxygen is abundant in swiftly flowing streams, but concentrations are lower in sluggish water and at the bottom of deep water.

To cope with these conditions, animals in mud bottom pools often exhibit adaptations, such as larger gills compared to stream animals. This adaptation allows them to extract more oxygen from the water. Examples of such adapted species include catfish and carp, showcasing how organisms in lotic ecosystems have evolved to thrive in diverse stream environments.

Some species of stonefly and mayfly naiads and caddisfly larvae are absent in cold water. Animals have various adaptations to light, temperature, oxygen, and feed. The density of an individual's biomass and productivity are less in sand bottom pools than in riffles and mud bottom pools. Clams and game fish, for example, are the inhabitants of sand bottom pools.

12.3.2.1.2 *Lentic ecosystem*

The lentic ecosystem is essentially the ecosystem of standing water although at times wind action stirs the surface layer and margins into great turbulence. For example, lakes, ponds, etc. are included in lentic ecosystems. The standing aquatic habitat offers a variety of unique conditions to which organisms must be adapted before they occur. Therefore, vertical stratification is an important aspect of standing water ecosystem in terms of light, temperature, nutrients, and oxygen availability. The aquatic lives tend to form distinctive vertical subcommunities in response to this stratification of physical factors.

1. Light: Light is essential to plants for photosynthesis and animals for feeding. Many organisms orient to light and some are sensitive to light. The vertical stratification of standing water regarding light is based on its penetration The vertical zone up to which the photosynthesis predominates is called the **trophogenic zone**. Below this zone there may be considerable photosynthesis but oxygen absorption is greater than oxygen release. This zone is called the **tropholytic zone**. Below this zone, there is no light called the **dark zone**, which is deep water.

2. Wind, Current, and Temperature: Wind and current are two important environmental factors, which can make some movement to the surface water. Similarly, temperature is also an important factor in regulating the water temperature. Based on wind, current, and temperature penetration, the water body horizontally is stratified as:

- **(i) Epilimnion**: It is the horizontal zone of the upper water body, where the water circulates and is fairly turbulent.
- **(ii) Hypolimnion**: This is the lower horizontal water body, which remains undisturbed by winds or currents.
- **(iii) Thermocline**: It is the zone of most rapid temperature decrease, generally involving a drop of at least 1 °C per meter of depth. Above this zone, temperature is usually uniform (Fig. 12.4).

The distribution of oxygen at various depths depends upon the presence or absence of a thermocline, the amount of vegetation, and the organic nature of the bottom.

3. Lentic Biocies: Lakes and ponds tend to undergo succession, which includes changes in the biological community as a response to an increase in nutrient levels. Based on physical characteristics, species composition, productivity, and distribution niches, there are two major lake communities called **oligotrophic** and **eutrophic lake** biocies.

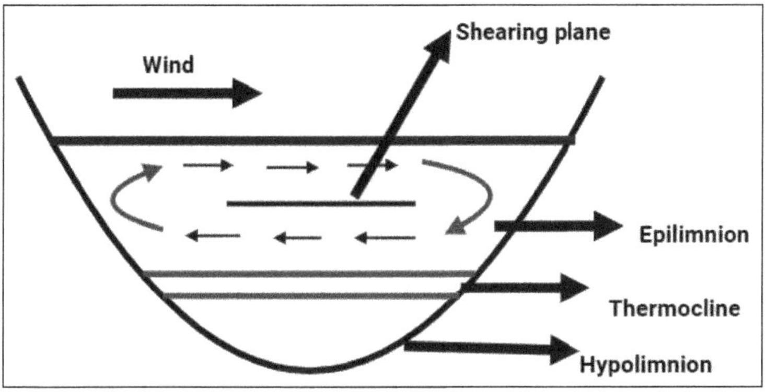

Figure 12.4 Horizontal stratification based on wind, current, and temperature.

Plankton consists of free-floating motile organisms, typically microscopic in size. Among plankton, those species that can be captured by a net are referred to as **net plankton**. **Nonoplankton** comprises organisms, such as bacteria, protozoa, and fungi, with sizes less than 0.06 mm in length.

Neuston refers to organisms that rely on the surface film as a substrate, while **Nekton** includes larger animals capable of independent locomotion, irrespective of water currents. In contrast, many animal species are bottom dwellers, attached to the substrate for support. These bottom-dwelling forms can be sessile (stationary), creeping, or burrowing, and include organisms like snails, burrowing worms, insect larvae, and bacteria. Collectively, these bottom-dwelling organisms are known as **benthos**.

The benthos can be divided into four subzones: (i) littoral zone, (ii) limnetic zone, (iii) profundal zone, and (iv) abyssal zone.

1. **Littoral Zone:** The littoral zone extends from the water's edge to the limit of rooted aquatic vegetation. This zone has abundant oxygen and good penetration of light. Littoral zone is further subdivided into **eulittoral** and **sublittoral zone**.

The **eulittoral zone** is the zone between high and low water wave action, i.e., the zone where the wave action is more active. In this zone, rooted vegetation is maximum. The **sublittoral zone** extends from the lower limit of the eulittoral zone to the lower limit of rooted vegetation. Beyond this, there is no vegetation.

2. **Limnetic Zone:** Limnetic zone is the open water zone away from shore. Vertically, it starts from the open surface water to the depth of light penetration, i.e., up to above the dark zone. Planktons are abundant in this zone.
3. **Profundal Zone:** Profundal zone is the deep-water zone below the littoral zone. This is the entire bottom below the rooted vegetation or biological community. It is commonly the bottom of the hypolimnion.
4. **Abyssal Zone:** This zone is found in deep lakes or oceans since it starts from about 2000 m from the surface. The water in this zone is very cold, there is no light penetration and low nutritional content.

12.3.2.2 Saline water ecosystem

The saline water ecosystem is the largest ecosystem, encompassing approximately 70% of the earth's surface. It includes oceans, seas, estuaries, and other saline environments. This ecosystem plays a crucial role in the global environment. Water, with its high heat capacity, can store significant amounts of energy. Oceans, representing vast and stable ecosystems, are commonly referred to as **marine ecosystems**. Some of the most significant marine ecosystems include:

- Oceans buffer and modulate the atmospheric climate.
- Oceans are major storehouses of carbon and exchange carbon dioxide rapidly with the atmosphere.
- It controls global warming.
- It is the greatest habitat for aquatic biota.
- Marine ecosystems regulate the atmospheric temperature and play a major role in hydrological cycles.
- Marine algae supply much of the earth's oxygen and take up a huge amount of atmospheric carbon dioxide.

- Oceans provide and produce food for many terrestrial organisms.

Marine ecology is concerned with environmental factors and problems of organismic adjustments quite different from those on land and different in many respects from those in freshwater.

Habitat

The marine ecosystem is considered to have the following divisions:

1. **Intertidal Zone (Littoral Zone)**: The intertidal or littoral zone extends between the limits of high and low tides. This biome is made up of areas exposed alternately to air during low tide and saline waters during high tide. The constant movement of water transports nutrients into and out of these areas. Large algae are found in these areas. Intertidal zones are major economic resources as shellfish and a variety of birds are present in this zone abundantly. The sublittoral zone covers the continental shelf to a depth of about 200 m.
2. **Pelagic Zone (or Open Water)**: The open waters of the oceans are called pelagic zone, which is away from the land. It is generally cold and there is thermal stratification with a constant mixing of warm and cold ocean currents. The pelagic zone is low in nutrients like nitrogen and phosphorus; thus, this is the zone with low productivity and low diversity of algae and other species.
3. **Benthic Zone (or Bottom Ocean)**: It starts below the pelagic zone but includes the deepest part of the ocean. The bottom of the zone consists of sand, salt, organic decays, etc. Temperature in this zone remains about 3–5 °C and it further decreases as we go further down in the ocean. The water is too dark for photosynthesis; thus no plants grow in this zone. The low density of fungi, bacteria, sponges, etc. is found in this zone.
4. **Abyssal Zone**: The deepest part of the ocean is known as abyssal zone. The water in this zone is very cold, there is no sunlight penetration, low nutritional content. Anaerobic bacteria can live in these low-oxygen bottom sediments. Below this abyssal zone, there are ocean trenches called hadal zone which extends much deeper.

Based on light penetration, the vertical stratification of the ocean includes the photic and aphotic zones. The **photic zone** is the vertical depth of the ocean up to which sunlight can penetrate and photosynthesis activities are carried out. The **aphotic zone** is the zone up to which there is no sunlight penetration and no photosynthesis activities. The **neritic** zone is above the continental shelf and the oceanic biosphere is further subdivided vertically as **epipelagic** and **mesopelagic** zones depending upon the penetration of light.

Marine Biocies

The marine biocies tend to form distinctive vertical sub-communities in response to the stratification of physical factors like light, oxygen, temperature, and nutrients. The marine life is generally distributed between the open ocean water ecosystem and the deep ocean water ecosystem (bottom).

Plankton

The plankton subcommunity consists mainly of microscopic plants, animals, and protists (single-celled organisms such as amoebae) that float freely within the water column. The **nanoplankton** consists of mostly protozoans, algae, bacteria, fungi, and viruses. The bacteria are largely **periphytic** in that they are attached to the surfaces of floating plants and animals and to particles of detritus.

The green **phytoplankton** is composed primarily of several kinds of algae. The algae are very important as they are the producers and represent the first trophic level of the oceanic food web or food chain.

The micro animals include zooplanktons which occupy the next trophic level of the oceanic food chain. The species composition of zooplankton is not uniform throughout the ocean.

Nektons

Nektons are larger animals capable of locomotion independent of water currents. For example, fishes range from sharks to numerous varieties of small species and mammals like seals, dolphins, and whales. The distribution of fishes is irregular, but in general, they occur more abundantly in neritic waters than in open oceans.

Benthos

Many animal species are bottom dwellers such as snails, burrowing worms, bacteria, etc. Benthos are of much greater variety and abundant in the littoral zone and decrease in numbers with depth. Oxygen levels are lower in the benthic environment. Anaerobic bacteria can live in these low-oxygen bottom sediments. Emergent plants such as cattails and rushes are rooted in the bottom sediments of the littoral zone (near shore). They create important structural and functional links between strata of the ecosystem and often are the greatest producers of net primary productivity in the ecosystem.

12.4 Evolutionary Ecology

The fascinating varieties of organisms and complex ecological relationships give the biosphere its unique productive characteristics. This leads to the biodiversity. Understanding of biological diversity is followed by the understanding of biological evolution, interactions between species, and the effects of environment on life diversity.

Charles Darwin and Alfred Wallace developed the theory of evolution. **Biological evolution** is a process through which new species are formed and it also refers to the change in inherited characteristics of a population from generation to generation through mutation, natural selection, migration, and genetic drifts.

The new species formed through biological evolution can no longer reproduce the original species. Thus, biological evolution is a one-way process. Once a species is extinct, it is gone forever.

1. Mutation: The gene, a complex chemical compound called DNA (deoxyribonucleic acid), is the building block of a life, which is inherited from one generation to the next. When a cell divides, the DNA is reproduced so that each new cell gets its share. Sometimes, there is a failure in the reproduction of the DNA due to the presence of toxic chemicals or other foreign bodies in it. Thus, it results in a change in the DNA and there is a change in inherited characteristics. The change of DNA characteristics by any means in the next generation is called **mutation**. Mutation can result in a new species giving them natural advantages so that they are relatively more likely to survive and reproduce and are better adapted than their parental species to the environment.

2. Natural Selection: Natural selection favors the survival of the fittest. It is a process by which the new species better adapted to a particular set of environmental conditions as compared to their less fit competitors. These new species through natural selection are also better adapted to the presence or deficiency of nutrients, diseases caused by bacteria or viruses, and temperature. Natural selection is relatively less effective in small populations than in large ones.

3. Migration: The migration of one population of a species into the habitat of other occupied species that can lead to changes in gene frequency is called hybridization. In hybridization, the result of introgression is the gradual intrusion of the characters of each population into the other, so that all distinction between them disappears. The hybrids show better adaptation than their parents. These hybrids will be selected for the replacement of both parental populations.

4. Genetic Drift: Genetic drift refers to changes in the frequency of a gene in a population because of chance rather than other processes of biological evolution. It is more prominent in small populations. When individuals are isolated in a small group from a larger population, there is a considerable decrease in the number of genes available to the main body of the species, as no individual or small group of individuals can possess all the genes that occur within the species pool. In this case, the individuals of small groups may not be better adapted to the environment, because the genetic variability of the species is greatly reduced. But in a large population, if the offspring of one pair of parents under-represents an allele, others will over-represent it. Thus, genetic drift is the establishment of restricted genotypes in small populations by drifting the allele (certain genes) frequencies away from their starting values. Successive fixation (random changes in allele frequency) events will lead to the progressive loss of genetic variation, often completely from the population.

5. Biodiversity: Biodiversity refers to the variety of organisms, number of organisms, and complex ecological relationships between organisms or groups of organisms and their environment. Biological diversity involves three important concepts such as:

(i) Genetic diversity refers to the number of genetic variations present in a particular habitat.

(ii) Species diversity describes the number of different kinds of organisms within an individual community or ecosystem.
(iii) Ecological diversity includes ecological niches, number of habitats, trophic levels, complexity of a biological community, and recycled materials within the ecosystem.

The confinement of a species to a specific ecological niche is influenced by its structural adaptations, physiological adjustments, tolerances, and behavioral patterns. An **ecological niche** represents a unique combination of physical factors (microhabitat) and biotic relations (role) that a species requires for its life activities and sustained existence within a community.

The term **"species"** refers to all organisms of the same kind that share sufficient similarities to naturally interbreed and produce viable offspring for the next generation. The coexistence of various species in distinct biological niches contributes to the **species diversity** within an ecosystem.

12.4.1 Interactions Between Species

Interactions may take place between individuals within and between species. These interactions are known as **population ecology**. The interactions may be interspecific or intraspecific. **Interspecific** interactions include mutualism, commensalism, and predation-parasitism. The **intraspecific** interactions include colonization, aggregation, and social organization

12.4.1.1 Interspecific interaction

1. Mutualism: Mutualism is an association between two or more species in which all participant species are benefitted. Sometimes the term **symbiosis** is used to define this relationship. For example, nitrogen-fixing bacteria with roots of legumes and lichens contain algae and fungi that require each other to persist.

2. Commensalism: Commensalism defines the interactions in which one species is benefitted and another species is neither benefitted nor harmed. For example, the saps drive the insects from grasses that are fed on extensively by other birds and many microorganisms take shelter in the canal system of sponges.

3. Predation-Parasitism: In this type of interaction, one species is benefitted and the other one is harmed. Parasitism is the relation between two individuals in which the parasite is benefitted and the host is harmed. In predation relations, one species kills the other.

12.4.1.2 Intraspecific interaction

1. Colonization: Colonization is the interaction of species within themselves to form colonies for survival. They form colonies for better protection from other predators or environmental conditions, improved utilization of food supplies, or more efficient reproduction. For example, the herd of elephants forms a colony in which the young ones are at the center and then the females. In the outer ring, the males are there to protect the younger elephants.

2. Aggregation: The aggregation of individuals must have survival value because it persists. For example, mayflies, midges, and mosquitoes swarm for mating purposes. In honey bees, when hive temperatures drop below 14 °C during winter days, they form clusters and maintain a mass temperature several degrees above the atmospheric temperature.

3. Social Organization: In this type of interaction, the species are well organized to protect, feed, and reproduce themselves for their survival. For example, the three primary organizations of ants are the winged reproductive males and females, the wingless sterile soldiers, and smaller wingless workers.

12.5 Biological Productivity

The rate of biomass production per unit area per unit time is called productivity. The total amount of organic matter in any ecosystem is called **biomass**. Biomass is increased through the biological production of growth, the changes in biomass over a given period of time are called **net production**. There are two kinds of production: (i) primary production and (ii) secondary production.

1. Primary Productivity: It is the rate of biomass production in a community. The primary productivity process is carried out by **autotrophs** by forming sugar from sunlight, carbon dioxide, and water. The energy left after respiration is net primary production.

(i) $6CO_2 + 6H_2O \rightarrow C_6H_{12}O_6 + 6O_2$ (Photosynthesis)

(ii) $CH_2O + 6O_2 \rightarrow 6CO_2 + 6H_2O + \text{Energy}$ (Respiration)

Net Production = Gross Production – Respiration (Photosynthesis)

2. Secondary Productivity: This process is carried out by the heterotrophs, which cannot make their food but must feed on other living organisms. The production by heterotrophs is called secondary production as it is derived from primary production. Secondary production is the rate of energy transferred and stored at consumer levels over a period of time (Fig. 12.5).

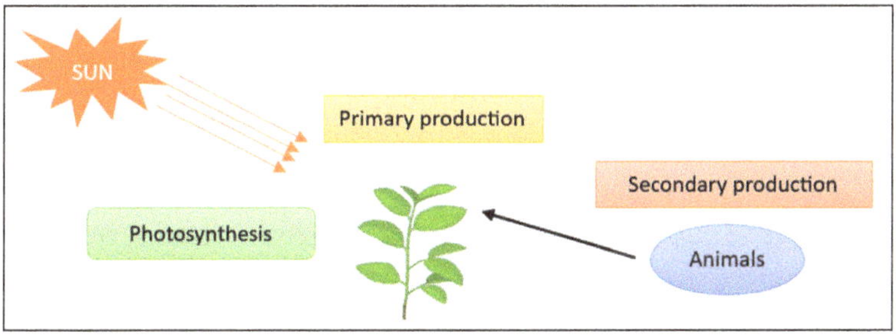

Figure 12.5 Primary and secondary productivity.

12.6 Energy Flow

Energy flow is the flow of energy through an ecosystem from the external source or environment (solar energy) through a series of organisms and back to the external environment (sink or outer space). Organisms use carbon dioxide, water, and sunlight and then return them to the environment in altered forms as byproducts of their metabolic processes. In the ecosystem, energy flows in a unidirectional path and eventually into some low-temperature sink such as outer space. For example, lions eat deer to get energy but deer cannot eat lions; thus, the flow of energy is unidirectional in an ecosystem.

The second law of thermodynamics states that energy always changes from a more useful and more highly organized form to a less useful and disorganized form. The ecosystem is said to be the intermediate between the energy source (sun) and the energy sink (outer space).

The energy efficiency flow in an ecosystem is commonly known as trophic level efficiency, which is the ratio of production of one trophic level to the production of the next lower trophic level, which is never high. The flow of energy starts from the green plants to the next trophic level and so on. Green plants can convert 1–3% of the energy received from the sun into plant energy. Then the herbivores convert the potentially available plant energy into herbivorous energy, which may be converted into carnivorous energy by carnivores. It is found that the transfer of total energy from one trophic level to another is only 10% of the gross productivity of producers. This is known as 10% low (Fig. 12.6). The energy flow in an ecosystem through a linked path is known as a food chain or food web.

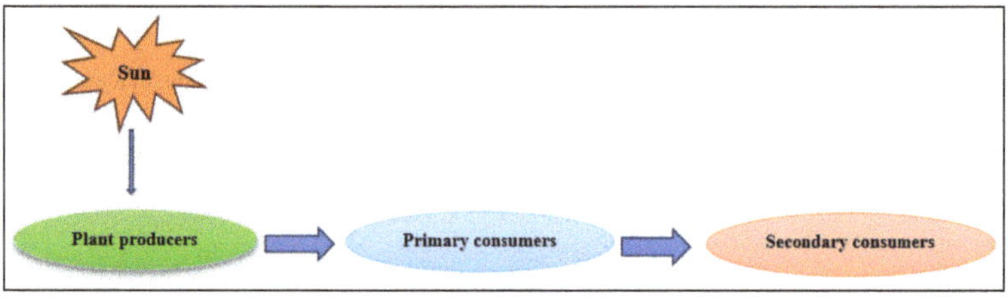

Figure 12.6 Energy flow.

12.7 Distribution of Toxicants in the Ecosystem

The various human activities and the advent of science and technology cause the release of harmful substances into the ecosystem. Out of these, certain toxic elements may be toxic in certain concentrations but are required in small quantities for metabolism. There are certain elements required in large quantities by certain organisms for their metabolic processes. Those elements required in small quantities are called micro-elements and elements required in large quantities are called macro-elements. Both macro and micro-elements are necessary for the growth and normal development of an organism. There are 9 macro-elements such as carbon (C), oxygen (O), hydrogen (H), nitrogen (N), phosphorus (P), sulfur (S), potassium (K), calcium (Ca), and magnesium (Mg). In addition, there are 17 micro-elements such as Na, Cl, Si, V, Cr, Mo, Mn, Fe, Co, Ni, Cu, Zn, B, Sn, Se, F, and I.

These substances are released into the environment and their behavior depends on many factors like physical, chemical, and/or biological.

The released harmful substances or toxicants can enter the ecosystem in any of the four matrices: (i) the atmosphere by evaporation, (ii) the lithosphere or soil by adsorption, (iii) the hydrosphere by dissolution, and (iv) the biosphere by absorption, inhalation, ingestion. In this section, we will emphasize the absorption of harmful substances by biota.

Various environmental transport processes contribute to the exposure of organisms to harmful substances and their subsequent uptake. In the environment, only a portion of the total quantity of a chemical present is potentially available for uptake by organisms, and this is referred to as bioavailability. The bioavailable fraction of toxic substances can either increase or decrease due to changes in factors such as solubility, chemical stability, and thermal stability of substances.

For instance, heavy metals can be released from the soil into surrounding or interstitial water through the activity of microorganisms. In soils, the bioavailability of toxic substances is controlled by sorption processes. This distribution is primarily determined by substance properties such as volatility, water solubility, and affinity for lipids and dead organic materials. The tight sorption of contaminants increases the residence time of the toxic substances in the soil. Slowly, these contaminants are released into other ecosystems, due to chemical action and/or microbial activities.

12.8 Terrestrial Ecotoxicology

An ecosystem comprises all types of organisms or groups of organisms that function together and interact with the physical environment. There is a correlation between different organisms and non-living components through energy flow and material cycles (Odum, 1983). The toxic substances can affect the ecosystems over a large area mainly because of a high degree of functional redundancy of the ecosystems. This can be explained as the organisms can take over each other's function through food chains, which run parallel to each other and are linked to form food webs in various ways. The movement of biotic and abiotic components within large ecosystems varies depending on several factors, including the species of animals and physical components of the surroundings.

Terrestrial ecotoxicology is restricted to the exposure and effects of toxic substances on terrestrial ecosystems only. Most terrestrial organisms are very mobile, covering a large area through migration and dispersion. In this way, all organisms contribute to the **diffuse stability** of an ecosystem.

Toxic substances can cause effects within terrestrial ecosystems at sub-organisms to super-organismic levels. For example, residues of DDT, other chlorinated hydrocarbon insecticides, and industrial chemicals including PCBs can influence molecules, tissues, organs, individuals, populations, and even whole communities. Terrestrial ecotoxicological tests must be performed at different levels of the organization, from molecules to ecosystems because dynamic population systems, interacting biological communities, or even the whole ecosystem makes it more complex. The toxic effects of chemicals on terrestrial organisms are observed in some avian species like Bermuda petrels, brown pelicans, etc. The problem of chemical effects on the avian population became more apparent as the relationship between DDT contamination and declining bird population was established. The study by Ratcliffe (1967) established the cause of declining of the peregrine falcon and sparrow hawks and the toxic substances (DDT) caused the eggshell thinning. These toxic elements also affect the reproductive systems of avian species resulting from eggshell thinning and decreased hatching success.

Ecosystems can experience imbalance due to the impacts of toxic elements. The selective action of toxicants affects different species in varying ways, to different extents, and at different concentrations. Toxic effects can be lethal, leading to the death of the species, or sublethal, where species remain alive but experience reduced developmental and reproductive abilities. It is important to note that not all toxicants directly cause harm; some may have secondary toxic effects.

For example, certain substances like nitrogen and phosphorus, which are essential plant nutrients, can have secondary toxic effects.

An excess of these nutrients in aquatic systems, often resulting from human activities such as fertilizer runoff, can lead to eutrophication. Eutrophication is a process where increased nutrient levels stimulate excessive plant and algae growth, disrupting the balance of the ecosystem and negatively impacting water quality. This is an indirect and undesirable consequence of the presence of substances that are not toxic by themselves but can lead to ecological imbalances when present in excess.

12.8.1 Lethal Effect

When toxicity is described in quantitative terms, the concept of **Lethal Dose** is used from where the lethal effect is derived. The toxic effect may be measured as the toxicological response level of individuals to exposure is calculated as the dose of a chemical substance that causes the death of a part of or the whole population.

The lethal dose and effect are said to be 100% when all the organisms die. It is known as LD_{100}. When 50% of the population dies after a certain period of exposure the lethal dose is LD_{50}. Lethal doses (LD) are usually included in **irreversible effects,** also known as lethal effects. When the effect of a toxicant in an organism is permanent, i.e., even if there is a complete elimination of toxic elements from the organism, still then the effect of toxicant exposure persists. This is known as an irreversible effect.

12.8.2 Sublethal Effect

The effective dose-50 (ED_{50}) is the dose that causes an effect in 50% of the population. Similarly, the toxic dose-50 (TD_{50}) is the dose that causes toxic effects on 50% of the population. In the case of ED_{50} and TD_{50}, the response is that one will die from the exposure to the toxicants. These two ED_{50} and TD_{50} show some positive responses, i.e., the ultimate effect is not the death of the organism.

In many cases, effective and toxic doses are detrimental and beneficial. When the effect from toxicant exposure is a temporary one and the effects are not observed after the elimination of toxic substances from the organism, the effect is said to be reversible or sublethal effects. Organisms exposed to toxicants with reversible effects do not die, these effects may be called sublethal. The sublethal doses of most toxic substances are eventually eliminated from an organism's system.

12.8.3 Non-lethal Effect

The distribution of animals within an ecosystem can be random, and certain toxicants may produce harmful effects on one type of living organism without affecting another, even if the two coexist in close proximity (Albert, 1973). Sensitivity to toxicants varies among populations, and it depends on how a population is regulated and the specific effects of the substance.

Two distinct types of effects are observed in populations: effects on reproduction and effects on predator-regulated populations. Predator-regulated populations tend to be more sensitive to toxicants affecting reproduction, while food-regulated populations are primarily susceptible to substances that influence survival.

As an example, organisms at the top of the food web (carnivores) are often food-regulated populations, meaning they are more sensitive to substances affecting survival. In contrast, organisms lower in the food web (herbivores) are typically predator-regulated populations, and they are more sensitive to toxicants that impact reproduction.

The organisms at the base of the food chain or food web are the most affected by the toxicant on reproduction because this is of great importance to the persistence of such populations. The carnivores will depend on the lower organisms for food. If the population of the lower organisms is reduced, then there will be a shortage of food for top carnivores and they will be affected for their survival.

12.9 Aquatic Ecosystem

Water is the largest sink for many chemicals from anthropogenic sources. An aquatic ecosystem is very important and complicated because most toxicants are easily soluble in water. Aquatic toxicology deals with the effects of toxicants on the living organisms in the aquatic environment. The major sources of toxicants in the aquatic environment are anthropogenic sources such as a group receiving the same dose of a particular toxicant and for another group, the dosage is gradually increased considering that all animals were equally sensitive to that toxicant. Then there would be a threshold dose below which all animals would remain alive without any major toxic effect is said to be a non-lethal effect.

12.10 Effects of Toxicants on Terrestrial Ecosystem

The structure and functioning of terrestrial ecosystems are so complicated that it is very difficult to predict the effects of toxic substances on populations, communities, and ecosystems. For example, the individuals within a population may show differences in sensitivity to a toxicant because of various factors like age, nutritional state, sex, body structure, and other physical factors. There are also genetic differences in sensitivity between individuals to certain toxicants.

In case of an essential element to living organisms display bimodality in terms of dose-response relationship. At a very low supply of an essential chemical element, the plant suffers from a shortage of nutrients, which may cause death to the plant. If the concentration of the same chemical element surpasses the optimum level for plant growth, it causes adverse effects that may result in death due to toxicity. The effects of toxicants on the terrestrial ecosystem are not primarily at the shortage site, but rather at the surplus site of the plant. Thus, the response of a plant exposed to a surplus of a chemical element is considered for the toxicity effects of industrial effluents, domestic and commercial sewages, etc., and terrestrial run-off and atmospheric deposition generally known as secondary pollutants. Many parameters are different from terrestrial ones that control the fate of the toxicants in the aquatic environment. Some important parameters are temperature, light, biological oxygen demand (BOD), chemical oxygen demand (COD), dissolved oxygen (DO), acidity, and alkalinity.

In terrestrial ecosystems, many toxicants undergo aerobic degradation, benefiting from the presence of sufficient oxygen. However, in aquatic systems, oxygen may not be readily available in significant amounts, resulting in limited aerobic degradation. This leads to the persistence of many toxicants for longer durations in aquatic ecosystems compared to terrestrial systems.

Aquatic animals are generally more susceptible to toxicants when compared to terrestrial animals. This increased susceptibility is attributed to the highly permeable nature of the skin and gills of aquatic organisms, which allows toxicants to easily enter their bodies. The temperature of the aquatic system plays a crucial role in controlling the metabolic activities of aquatic organisms, including fishes, invertebrates, and amphibians. Additionally, water temperature influences the solubility of toxicants in the aquatic environment. As a result, the accumulation and penetration of toxicants in aquatic organisms are significantly influenced by water temperature.

The other important factor in the aquatic ecosystem is pH value because most metal toxicants or metal-bound toxicants are easily soluble in acidic conditions, i.e., at a very low value of pH, most metals are soluble in water.

Acid deposition may be stored up in the snow and released during the summer days when the snow melts resulting in an increase in the acidity of the water. Thus, more soluble metals are available to aquatic organisms.

12.11 Genotoxicity

Genotoxicity is the study of the toxic effects of toxicants like mutagens, clastogens, aneugens, and teratogens on the genetic material of living organisms was originated with the experiments of Miller (1926). Life and propagation of life are determined by the information stored in the genes, which is carried by deoxyribonucleic acid (DNA). The genotoxic compounds, either directly or indirectly (biotransformation) can bind to the DNA molecule in different ways. The most important attacking sites are the nucleotide bases, which lead to mutagenicity.

Mutagenicity is the process by which the genetic information for determining and maintaining the integrity, functions, and relationships of living cells is permanently changed. Such effects may be observed as DNA strand breaks, base modifications, chromosomal rearrangements, fragmentation, and/or aneuploidy.

The primary target for mutagenic damage is DNA, and this damage is not limited to nuclear chromosomes. It can also occur outside the nucleus in mitochondria and the chloroplasts of plants. When toxic effects occur in germinal tissues, it can result in heritable effects, leading to an increase in genetic load. Mutations, alterations in the DNA sequence, can arise spontaneously, but they can also be induced by physical and chemical factors.

Human activities release a substantial amount of toxicants into the environment, contributing to a significant increase in mutation pressure. The mutagenic pressure, or the likelihood of mutations occurring, depends greatly on the timing of exposure to mutation-inducing factors. If these factors act on germinal tissues, they can affect the genetic material that is passed on to future generations, potentially leading to heritable mutations and an increased genetic load in populations. Understanding and mitigating mutagenic pressure is crucial for preserving genetic integrity and biodiversity in ecosystems.

Chromosomal damage is defined as the modification of the number or structure of chromosomes. Variations in the number of chromosomes may involve the complete complement (polyploidy) or only some of the chromosomes (aneuploidy). The interaction between the bases of the DNA or electrophilic substances may form covalent bonds with nucleophilic bases, thus, leading to DNA abducts, etc. DNA adduct formation is the genotoxic endpoint. Non-lethal genetic alterations in dividing somatic cells may ultimately result in the development of cancer. The DNA abducts can result in mispairing during subsequent replication or clastogenic changes such as strand breaks, which are thought to play a key role in the initiation, early propagation, and later stages of tumor progression.

12.11.1 Types of Mutation

There are three groups of mutation observed as follows:

1. Gene Mutations: Gene mutations are small changes in the bases of DNA, which are invisible in the light microscope. These are also called single-point mutations. These single-point mutations can be based on changes in the sequence of nitrogen bases, although phosphate oxygen is also a target. These changes occur due to the addition or deletion of bases. Base change may result in the coding of a different amino acid. This type of change is a miss-sense mutation.

2. Structural Chromosomal Aberrations: Chemicals or toxicants that cause structural chromosomal aberrations are also known as clastogenes. **Clastogenesis** refers to the process resulting in additions, deletions, or rearrangements of parts of the chromosomes that are detectable by a light microscope. Deletion or insertion of bases in a gene may be large or small. Chromatid gaps that are achromatic lesions in a chromosome may vary in length and are thought to be due to loss of DNA. Similarly, **breaks** are broken ends of chromatids that are dislocated but still contained within the metaphase. **Terminal deletions** occur when the chromosomal materials are spilt-off at the end of both chromatids at homologous places resulting in the formation of two chromatid fragments. In **centric ring chromosomes**, chromatid segments are centromeres and joined to form a ring. The ring chromosomes without centromere are called **acentric chromosomes**. In **chromatid exchanges,** the chromosome breaks and exchanges between chromatids, which can give rise to symmetrical or/and asymmetrical shapes.

3. Genome Mutations: Mutations that lead to abnormalities in the number of chromosomes are known as **aneuploidy**. The chemicals or toxicants responsible for causing aneuploidy are referred to as **aneugens**. Genome mutations resulting in aneuploidy occur due to unequal division of chromosomes, leading to cells with either more or fewer chromosomes than normal. Variations in the number of chromosomes can involve multiples of the complete chromosomal complement, known as **polyploidy**, or the elimination or multiplication of only some of the chromosomes, which is termed aneuploidy. Aneugens induce aneuploidy by interfering with the normal process of chromosomal division. For example, colchicine is a chemical that induces polyploidy by halting mitosis at metaphase, resulting in cells with a doubled quantity of chromosomal material and leading to a polyploid cell. Other compounds that induce polyploidy include vinca alkaloids such as vincristine, vinblastine, and podophyllotoxin.

12.11.2 Teratogenesis

Teratogenesis is the science of birth defects due to structural malformation caused by radiation, viruses, and toxicants, including drugs (Imlay et al., 1988). The agent that when administered prenatally, induces permanent structural malformations or defects in the offspring are called **teratogens**. Teratogens affect developing embryos adversely when exposed. Structural malformation occurs due to damage of embryonic or fetal cells.

Chapter 13

Biomarkers and Environmental Monitoring

"The key to expose environmental devastation"

13.1 Introduction

Environmental monitoring is an approach to evaluate environmental conditions and trends, support policy development and its implementation, and develop information for reporting to national policymakers, international forums, and the public. Biomarkers are important tools to estimate and monitor the bioavailability of contaminants in the environment integrated over a specific period. The main objectives of this chapter are to provide an overview of studies that use biomarkers for the assessment of exposure and the toxic impact of environmental contaminants on various organisms.

The environment is continuously loaded with foreign chemical substances, released by anthropogenic activities, and contributes positively to human society, e.g., by providing food and welfare, there are also numerous examples of chemicals that have caused harm to human health as well as the ecosystem. Environmental monitoring is used in the preparation of environmental impact assessments, as well as in many circumstances in which human activities carry a risk of harmful effects on the natural environment. It provides information

Environmental Toxicology: Concepts and Principles
Mohd Salim Reshi, Darakhshan Javaid, Syed Sanober Qadri, and Neelu Sinha

ISBN 978-981-5129-68-7 (Hardcover), 978-1-003-63707-3 (eBook)
www.jennystanford.com

about the distribution of chemicals in the environment and their temporal trends. It is the surveillance of environmental parameters like air, water, noise, soil, etc. to describe their quality. Biomarkers of environmental toxicants are measures of exposures and effects, which assess disease risk and inter-individual susceptibilities. They have successfully been used in environmental monitoring and assessment around the world to detect exposure to and effects of chemicals, metabolites, proteins, and DNA adducts to explicate mechanisms of bioactivation and detoxification of reactive toxicant intermediates. Here are key concepts related to environmental toxicity and biomonitoring.

13.2 Environmental Toxicity

1. Sources of Pollution: Pollution can come from various sources such as industrial activities, agriculture, transportation, and improper waste disposal.

2. Types of Pollutants: Pollutants can be chemical substances, heavy metals, pesticides, persistent organic pollutants (POPs), and other contaminants that can accumulate in air, water, soil, and food. It is somewhat artificial to consider air, water, and soil pollution separately because their effects are interchangeable. Chemicals emitted into the air eventually combine with rain or snow and settle down to become water and land pollutants. On the other hand, volatile chemicals from soil or those that enter lakes and rivers evaporate to become air pollutants.

Pesticides sprayed on land are carried by the wind to become transient air pollutants that eventually settle somewhere on land or water. For discussion purposes, however, some systematic division appears to be advisable.

13.3 Biomonitoring

Definition: Biomarkers are defined as changes in biological responses (ranging from molecular through cellular, and from physiological responses to behavioral changes) that can be related to exposure to, or to the toxic effects of, environmental chemicals.

Biological Markers: These markers can include chemicals, metabolites, or other substances that are indicative of exposure to environmental pollutants. For example, measuring levels of certain chemicals in blood or urine. These are useful tools for understanding the nature and extent of human exposure and risk from environmental toxicants (Fig. 13.1).

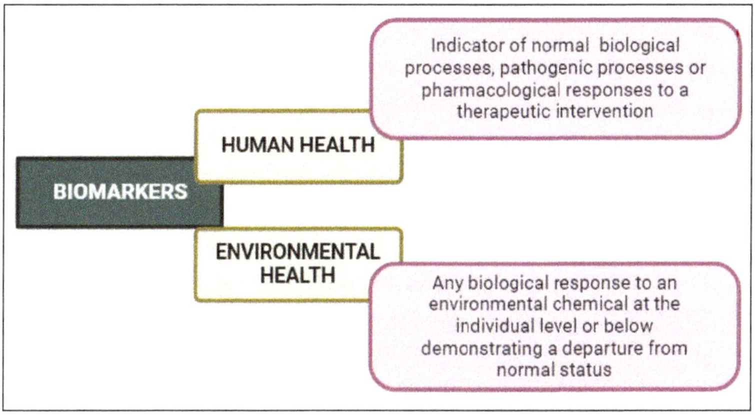

Figure 13.1 Role of biomarkers in human and environmental health.

They can serve as quantitative measures of chemical exposures and biologically effective doses, as well as early warning signals of biological effects. They can help increase the understanding of the processes by which a chemical is transported and transformed within an organism to produce a dose to a target tissue and the interactions at the cellular and molecular levels leading to a toxic endpoint. Additionally, biological differences or genetic polymorphisms may exist that cause some individuals to be more susceptible to environmentally induced diseases and serve as markers of susceptibility. It is useful to envision the processes that link exposure, dose, and effect as a continuum, as shown in Fig. 13.2.

Human Biomonitoring: In the context of human health, biomonitoring can be used to assess the levels of pollutants in the human body and estimate potential health risks.

Wildlife Biomonitoring: Similar approaches can be applied to assess the impact of environmental pollution on wildlife by studying specific biomarkers in their tissues or fluids.

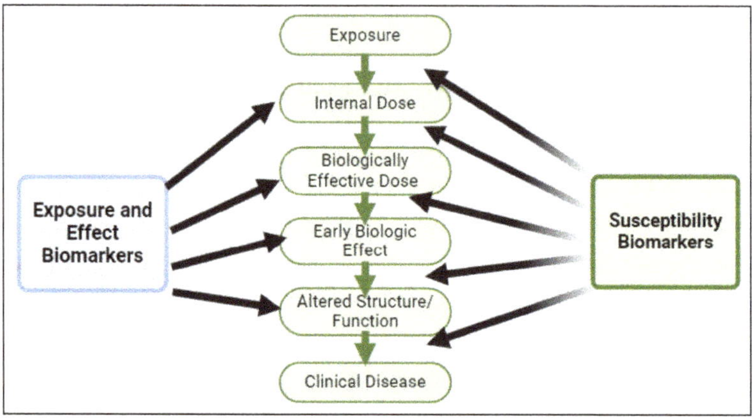

Figure 13.2 Exposure, effects, and susceptibility of biomarkers.

Biomonitoring involves the measurement of specific biological markers or indicators to assess the exposure of living organisms to environmental pollutants. Some environmental chemicals act as modulators of gene and protein activity and induce dysbiosis of the microbiome, which impacts the metabolome and overall health. Many different biomarkers have been used or are under development, and they all have different strengths and weaknesses. All monitoring strategies have reasons and justifications which are often designed to establish the current status of an environment and prediction of the impact of future development and/or alteration in the operation and design of existing installations. The use of biomonitoring methods in the control strategies for chemical pollution has advantages over chemical monitoring.

First, these methods evaluate effects in which the bioavailability of the compounds of interest is integrated with the concentration of the compounds and their intrinsic toxicity.

Second, most biological measurements form the only way of integrating the effects on many individual and interactive processes.

It has been demonstrated that biomarkers are useful monitoring tools, but further information is needed about the relation between the response of biomarkers and the health and fitness of organisms.

Although the field of biomarkers is still relatively new, many different analytical techniques have been developed to quantify such events as exposure to a certain chemical or early biological events resulting from exposures. Problems exist with many current biomarkers in that they have not been validated for use in large population studies and their significance for predicting the risk of clinical disease is unknown. In other words, although a marker may indicate exposure to a certain environmental chemical, it may not be well understood how that marker relates to other events in the toxicological paradigm. For example, does the marker of exposure relate to an exposure dose, an internal dose, a target organ dose, or a biologically effective dose? To what types of effects, such as early biological effects and disease endpoints, can this biomarker be linked? Additionally, what does this mean in terms of quantifying the risk of an adverse effect identifying susceptible groups, or even in terms of identifying early disease stages? Ideally, a series of biomarkers can be used to help strengthen the knowledge base about the entire disease continuum, from exposure to effect, or disease outcome. One example where biomarkers have been successfully used in risk assessments involves exposure to aflatoxin. Aflatoxin is a hepatocarcinogen produced by fungi that can contaminate certain foods, such as corn and peanuts. People are exposed to aflatoxin when they eat these contaminated foods. Aflatoxin is metabolized in the liver by CYP450 isoforms to form a highly reactive intermediate, AFB1-8,9-epoxide, which can bind covalently to DNA and proteins. Aflatoxin-albumin adducts in serum are one of the most reliable biomarkers of recent exposure. AFB1-8,9-epoxide can form DNA adducts through covalent binding to the N7 atom of guanine, yielding G-T transversion mutations. Aflatoxin-N7-guanine adducts in urine are a biomarker of biologically effective dose, and aflatoxin-N7-guanine adducts in the liver are a biomarker of early biological effect. DNA adduct levels in urine have been associated with measures of exposure and cancer incidence in experimental studies and large population studies.

Sample Collection: Collection of samples such as blood, urine, hair, or tissues from individuals or populations.

Analytical Techniques: Various laboratory techniques, including mass spectrometry, chromatography, and immunoassays, are used to analyze the collected samples for specific biomarkers.

13.3.1 Validation of Biomarkers

To evaluate the use of biological measurement as a biomarker, one must understand the relationship between the marker and the condition/disease of interest. Sensitivity and specificity are both critical components of the evaluation process. Sensitivity refers to the ability of a measurement to detect positive responses; whereas, specificity refers to the ability of a measurement to identify negative responses (to limit the number of false positives). Since one of the primary purposes of biomarkers in environmental health research is to identify highly exposed groups to predict or prevent disease, the biomarkers must not only be evaluated for their ability to assess the presence or absence of exposure or disease, but also for their ability to quantify the exposure, dose, or level of disease.

The evaluation of biomarkers includes the "backward" process of associating the marker with an exposure and the "forward" process of linking the biomarker with an effect that depends on its anticipated use. A biomarker observed before the onset of disease may have a low predictive value as a biomarker of effect but may be very useful as a marker of exposure, allowing long-term monitoring of an exposed population. On the other hand, a biomarker of effect that is expressed long after the exposure could be of relatively little use in exposure assessment, but be very useful in predicting the progression of disease or assessing risk. Animal models are useful in understanding the mechanistic basis of the expression of markers and the relationship between exposure, early effects, and disease. The validity of a biomarker of effect depends on the reliability of studies that provide the background data, particularly on mechanisms. Estimates of the sensitivity of a biomarker should include its evaluation in an unexposed population or unexposed animals to determine a baseline value for the marker. This evaluation may be difficult in pediatric populations due to ethical issues, such as invasive procedures with little benefit for the participant. Multiple biomarkers that can fill knowledge gaps across different points of the exposure-dose-effect continuum and/or that can be applied in a clinical setting.

1. Animal or epidemiology studies: These studies explore the relationship between biomarkers of exposure and measures of subclinical disease (early biological effect or altered structure/

function). Additionally, these studies could be expanded to explore the relationship between the subclinical disease measure and the actual clinical disease.

2. Mechanistic studies: These studies (e.g., using genomics or proteomics) of toxicant response linked to clinical disease. For example, the identification of the functional relevance of proteins where genetic polymorphisms have been found to modify the effect of environmental exposure on a disease endpoint.

3. Studies to validate the utility of biomarkers: These are used in large population studies, e.g., reliability, predictive value, sensitivity, specificity, affordability, applicability to the general population, and susceptible subpopulations.

13.4 Biomarkers of Environmental Toxicants Are Measures of Exposures and Effects

The gut microbiome has emerged as a new mediator in human health. Functional gut microbiome alterations that may adversely influence human health are an increasingly appreciated mechanism by which environmental chemicals exert their toxic effects. The functional damage driven by environmental exposure to the gut microbiome is gut microbiome toxicity.

Living organisms respond to environmental changes and xenobiotic exposures via diverse mechanisms. tRNA-mediated mechanisms are only recently emerging as important modulators of cellular stress responses. Through many ways, nucleoside modifications confer high functional diversity to tRNAs, with a focus on tRNA modification-mediated regulation of the eukaryotic response to environmental stress and toxicant exposures.

Exposure to heavy metals is ubiquitous and has been associated with several human diseases. Biomarkers of heavy metal exposure in children are particularly important for monitoring exposure and health risk assessment.

Benzene is a known carcinogen and causes hematotoxicity. Benzene exposure occurs through factory occupations and from emissions from burning coal and oil in the air. Cigarette smoking is another important source of exposure to benzene. Tranfo and colleagues employed a targeted LC-MS/MS to bio-monitor benzene through the biomarker S-phenyl-mercapturic acid (SPMA), which is formed from the benzene oxide metabolite conjugated with glutathione (GSH), and then excreted in urine as SPMA. The findings reveal that the main source of benzene exposure in a cohort not occupationally exposed to benzene in central Italy was active smoking; however, non-smokers were also exposed to airborne concentrations of this carcinogen.

Phthalates are used as plasticizers and additives in many consumer products. Human exposure to phthalates is prevalent and occurs mainly through dietary sources, dermal absorption, and inhalation. Laboratory animal studies reveal endocrine-disrupting and reproductive effects of phthalates. Thus, human exposure to phthalates is a public health concern.

Aldehydes are ubiquitous in the environment, originating from man-made sources, tobacco smoke, and natural processes. Some aldehydes are implicated in diseases, including diabetes, cardiovascular diseases, neurodegenerative disorders (i.e., Alzheimer's and Parkinson's Diseases), and cancer. Aldehydes are strong electrophiles that react with nucleophilic sites in DNA and proteins to form reversible and irreversible modifications. These modifications, if not eliminated or repaired, can lead to alteration in cellular homeostasis and cell death and contribute to disease pathogenesis.

Hemoglobin (Hb) and albumin (Alb) are the most abundant proteins in the blood and form covalent adducts with toxicants and endogenous electrophiles.

Aristolochic acids (AAs) are found in Aristolochia plants, some of which have been used in the preparation of traditional herbal medicines worldwide. AAs are highly nephrotoxic and carcinogenic to humans and are implicated as causative agents of the Balkan endemic nephropathy (BEN) and "Chinese herbs nephropathy" in Asia. The exposure of AAs in the Balkan Peninsula, where the comingling of Aristolochia plants with grains and the release of AAs from decayed seeds of Aristolochia plants contaminate the agricultural soil, the food crops, and the water supply. The links between exposure to AAs and their biomarkers of DNA damage, mutations in cancer driver genes, and mechanisms of kidney fibrosis in Asian cohorts are reported.

Data Interpretation: The results obtained from biomonitoring are interpreted to assess the extent of exposure, identify trends over time, and evaluate potential health risks.

13.5 Applications

1. **Human Health Risk Assessment:** Biomonitoring helps in understanding the levels of exposure to environmental pollutants and assessing potential health risks in human populations.

2. **Environmental Impact Assessment:** Biomonitoring in wildlife can provide insights into the impact of pollutants on ecosystems and biodiversity.

3. **Policy and Regulation:** Information from biomonitoring studies can inform the development of environmental policies and regulations to mitigate pollution and protect human and environmental health.

Biomonitoring is a valuable tool for understanding the complex interactions between the environment and living organisms. It plays a crucial role in assessing the impact of environmental toxicity on health and ecosystems and guiding efforts to mitigate these effects.

13.6 Case Study

Exposure to heavy metals is ubiquitous and has been associated with several human diseases. Biomarkers of heavy metal exposure in children are particularly important for monitoring exposure and health risk assessment. The study by Jursa et al. measured hair Mn, Pb, Cd, and As levels in children from the Mid-Ohio Valley to determine within and between subject predictors of hair metal levels. Specifically, occipital scalp hair was collected between 2009 and 2010 from 222 children aged 6–12 years (169 females, 53 males) participating in a study of chemical exposure and neurodevelopment in an industrial region of the Mid-Ohio Valley. They found that hair Mn and Pb levels were comparable (median 0.11 μg/g and 0.15 μg/g, respectively) and were ~10-fold higher than hair Cd and as levels (0.007 μg/g and 0.018 μg/g, respectively). In addition, metal levels were different between male and female subjects and showed different profiles along hair segments.

13.7 Conclusion

The main objective of this chapter is to provide an overview of studies that use biomarkers for the assessment of exposure and the toxic impact of environmental contaminants on various organisms.

Environmental toxicity refers to the harmful effects that various environmental agents or pollutants can have on living organisms, including humans, animals, and plants. These agents may include chemicals, heavy metals, pesticides, air pollutants, water contaminants, and other substances that, when present in the environment, can negatively impact the health of organisms. Biomonitoring is a scientific process used to assess and measure the exposure of living organisms to these environmental pollutants by analyzing specific biological markers or indicators. With respect to future biomarker research, it is important to realize that different concepts are needed for the specific purposes of environmental monitoring programs. In conclusion, it can be stated that biomarkers are promising tools for environmental risk assessment (ERA). In view of the present chemically oriented pollution-abatement policies and the need to reveal specific chemical problems, it is most probable that biological-effect analysis will never totally replace chemical analyses. The biomarker approach, therefore, should not be considered as a replacement for conventional assessment techniques, but as an important supplementary approach of great ecological relevance. Several guidelines for ecotoxicological research leading to an actual incorporation of biomarkers in ERA monitoring are proposed in this chapter. Much work must be done to test and interpret biomarker responses and to develop acceptable quality assurance (QA) procedures. Only when both scientific and legal credibility of this information is established can the biomarker techniques be fully applied in routine monitoring programs. Since it seems obvious that chemical monitoring alone is insufficient for a reliable classification of water quality, the efforts to incorporate biological compounds into the ERA research will eventually be worthwhile.

Chapter 14

Role of Environmental Toxicology in Public Health

"Defending public health from unseen hazards"

14.1 Introduction

The role of environmental toxicology in public health revolves around identifying and characterizing environmental health hazards. From air pollutants and pesticides to heavy metals and industrial chemicals, potential hazards surround us in our daily lives. Environmental toxicology is a multidisciplinary domain of science that occupies an important niche, overlapping the fields of toxicology, environmental health, and public policy. While studying the adverse health effects of chemical, biological, and physical agents on living organisms in the ecosystems, environmental toxicology focuses on humans, and therefore, plays an important role in addressing public health challenges. Over the years, environmental toxicology research has been providing tools and scientific evidence to policymakers and the public to prevent substantially greater environmental degradation, including adverse human health impacts. By understanding the toxicological profiles, researchers and public health officials assess

Environmental Toxicology: Concepts and Principles
Mohd Salim Reshi, Darakhshan Javaid, Syed Sanober Qadri, and Neelu Sinha

ISBN 978-981-5129-68-7 (Hardcover), 978-1-003-63707-3 (eBook)
www.jennystanford.com

the risks to different population groups. Children, pregnant women, and the elderly may be particularly vulnerable to certain toxins, necessitating tailored risk assessments and protective measures.

Unfortunately, in most developing countries, despite the existence of abundant resources for the release of pollutants into the environment and on the other side higher incidence of human health problems, there is not enough effort for management of environmental health maintenance. This might be due to a lack of attention to environmental toxicology issues associated with the health impact of pollutants.

One of the primary functions of environmental toxicology is to identify and assess environmental hazards. It investigates pollutants, such as heavy metals, pesticides, air and water contaminants, and their potential health impacts. By determining the sources, exposure pathways, and levels of these toxicants, researchers can quantify the risks to public health. This information forms the basis for developing mitigation strategies and public health interventions. Environmental toxicologists play a key role in the risk assessment process. They analyze exposure data, toxicity profiles of chemicals, and the vulnerability of various population groups. This data is used to set safe exposure limits and guidelines, helping governments and regulatory bodies establish regulations to protect public health. Risk management strategies are designed to reduce exposure to toxicants, whether by stricter regulations, improved pollution control, or public education campaigns. Environmental toxicology contributes to public health surveillance by monitoring the presence of toxicants in air, water, soil, and food. Through these surveillance programs, we can track changes in environmental quality and detect trends in chemical exposure. Early detection of emerging threats allows for swift action to minimize the risk and protect public health.

14.2 Risk Assessment

It is a scientific process used by federal agencies and risk management decision-makers to make informed decisions about actions that may be taken to protect human health by ascertaining potential human health risks or health hazards associated with exposure to chemicals in the environment. Some real-world examples of risk assessment include establishment of national ambient air quality and drinking water standards for protection of public health (e.g., ozone, particulate matter in outdoor air; chromium, chloroform, or benzene in water); establishment of clean-up levels for hazardous waste site remediation; development of fish consumption advisories for pregnant women and general population (e.g., PCBs, mercury); assessment of risks and benefits of different alternative fuels for sound energy policy development (e.g., oxygenated gasoline, biodiesel); and estimation of health risks associated with pesticide residues in food. The estimated risk is a function of exposure and toxicity. The regulatory risk assessment follows a four-step paradigm using qualitative and/or quantitative approaches. The four steps of risk assessment are as follows:

1. Hazard assessment
2. Dose–response assessment
3. Exposure assessment
4. Risk characterization

The purpose of risk assessment is to evaluate the severity of harmful effects on human health and the environment that may result from exposure to chemicals present in the environment. The Environmental Protection Agency (EPA) procedure of risk assessment, whether related to human health or to the environment (Fig. 14.1) involves four steps:

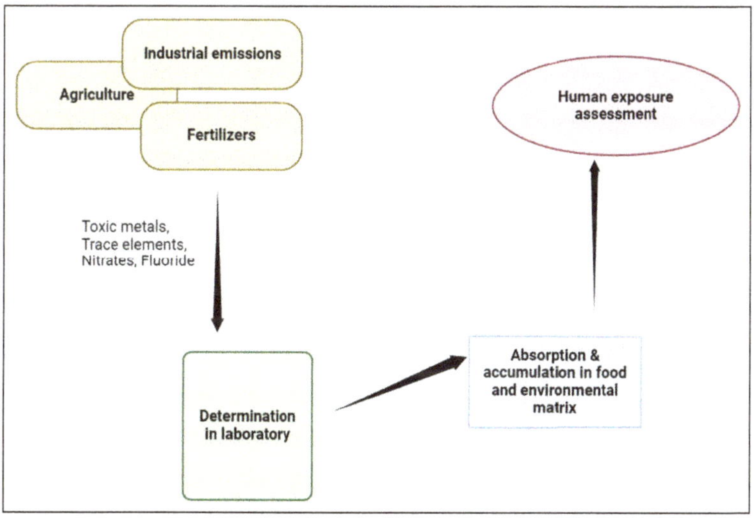

Figure 14.1 Environmental toxicology.

Step 1: Hazard identification is the first step used in a process to assess risk. The result of a hazard analysis is the identification of different types of hazards. A hazard is a potential condition and exists or not (probability is 1 or 0). It may in single existence or in combination with other hazards (sometimes called events) and conditions become an actual functional failure or accident (Mishap). Hazard Identification and Risk Assessment is a method, by which, we try to identify the main hazardous substance, and then try to reduce the effect of the hazard. In this step, a scientific weight of evidence analysis is performed to determine whether a particular substance or chemical is or is not causally linked to any health effect at environmentally relevant concentrations. For this purpose, we will use various risk assessment methodologies by knowing the hazard consequences and by analyzing all the processes that are being carried out during handling. This is performed to establish whether, and to what degree, toxic effects in one setting will occur in other settings. The evidence comes from human and animal studies.

Hazard identification involves identifying the contaminants that are suspected to pose health hazards, quantifying the concentrations at which they are present in the environment, describing the specific forms of toxicity (neurotoxicity, carcinogenicity, etc.) that the contaminants of concern can cause, and evaluating the conditions under which these forms of toxicity might be expressed in exposed humans.

Step 2: Toxicity or dose-response assessment takes the toxicity data gathered in the hazard identification step from animal studies and exposed human population studies and describes the quantitative relationship between the amount of exposure to a chemical (or dose) and the extent of toxic injury or disease (or response). It involves further evaluating the conditions under which the toxic properties of a chemical might be confirmed in exposed people, with particular emphasis on the quantitative relation between the dose and the toxic response. Generally, as the dose of a chemical increases, the toxic response increases either in the severity of the injury or in the incidence of response in the affected population. The development of this relationship may involve the use of mathematical models. This step may include an assessment of variations in response, e.g., differences in susceptibility between young and old people.

Step 3: Exposure assessment involves specifying the population that might be exposed to the agent of concern, identifying the routes through which exposure can occur, and estimating the magnitude, duration, and timing of the doses that people might receive because of their exposure. It determines the magnitude of the amount of an agent to which humans are exposed (i.e., exposure concentration), and the amount of dose (or intake) is estimated by taking the magnitude, frequency, duration, and route of exposure into account. Exposure assessments may consider past, present, and future exposures.

Step 4: Risk characterization involves integrating information from the first three steps to develop a qualitative or quantitative estimate of the likelihood that any hazard associated with the agent of concern will be realized in exposed people. This is the step in which risk-assessment results are expressed. Risk characterization should also include a full discussion of the uncertainties associated with the estimates of risk (Adapted from NRC 1994). Based on information obtained from the risk assessment, decisions are made about the best way to address environmental contamination and exposure. The risk manager also includes an evaluation of social, legal, economic, and policy issues to determine the best approach to address an exposure issue. ATSDR's public health assessment is an evaluation of environmental data, health outcome data, and community concerns associated with a site where hazardous substances have been released. The health assessment identifies populations living or working on or near hazardous waste sites for which more actions or studies are needed.

The improvement in the scientific quality and validity of health risk estimates depends on advancements in our understanding of human exposure to and toxic effects associated with, chemicals present in environmental and occupational settings. Risk assessments are important for informed regulatory decision-making in environmental sustainability and to ensure that costs associated with different technological alternatives are scientifically justified and protest public health. Although significant uncertainties remain, this risk assessment methodology has been extensively peer-reviewed, is widely used, and understood by the scientific community, and continues to expand and evolve as scientific knowledge advances.

Risk management (Fig. 14.2) is distinct from risk assessment, and it involves the integration of risk assessment with other considerations, such as economic, social, or legal concerns, to reach regulatory decisions regarding the need for and practicability of implementing various risk reduction activities.

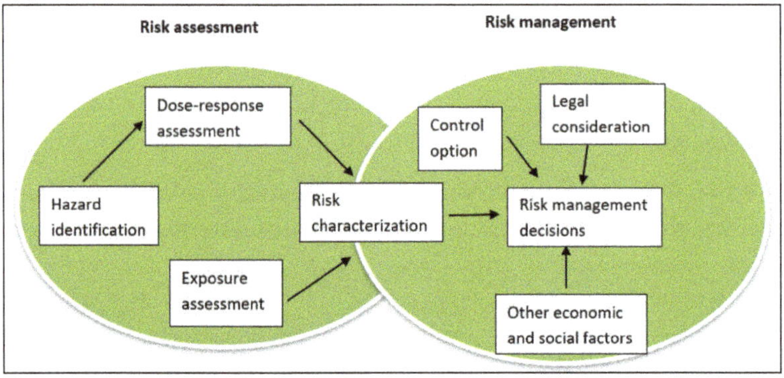

Figure 14.2 Risk assessment vs risk management.

14.3 Case Studies

The ATSDR case studies included in this chapter are original copies of the ATSDR cases and can be obtained from: Continuing Education Coordinator, ATSDR, Division of Health Education (E33), 1600 Clifton Road, N.E., Atlanta, GA 30333.

14.3.1 A 35-Year-Old Carpenter with Peripheral Neuropathy and Skin Lesions (Arsenic Toxicity)

A 35-year-old, fair-skinned male began approximately 3 months ago, with the symptoms of insidious onset of numbness and tingling in his toes and fingertips, progressing slowly in the ensuing weeks to involve the feet and hands in a symmetric "stocking-glove" fashion. In the past 2–3 weeks, the tingling has taken on a progressively painful, burning quality, and he has noted weakness in gripping tools. No ataxia, dysphagia, visual symptoms, or bowel or bladder incontinence have been noted, and he has not complained of headaches, back or neck pain, or confusion. The patient's medical history is remarkable for a flu-like illness approximately 4 months ago characterized by 3–4 days of fever, cough, diarrhea, and myalgias, which resolved spontaneously.

Further questioning revealed that the patient had been a carpenter since completing high school 17 years ago. For the last 10 years, he has lived in a rural, wooded area in a home he built. Neurologic examination reveals diminished proprioception in the hands and feet, with a hyperesthetic response to pinprick sensation on the soles. Motor bulk and tone are normal, but there is slight bilateral muscular weakness in the dorsiflexors' toes and ankles, wrist extensors, and hand intrinsics. Reflexes are absent at the ankles, biceps, and knees. Coordination and cranial nerve function are within normal limits. Dermatologic examination reveals brown patches of hyperpigmentation, with scattered overlying pale spots in and around the axillae, groin, nipples, and neck. The palms and soles show multiple hyperkeratotic corn-like elevations 4–10 mm in diameter. Three irregularly shaped, sharply demarcated, erythematous, and scaly plaques, measuring 2–3 cm, are noted on the patient's torso. The remainder of the physical examination is normal. On initial laboratory evaluation, the CBC shows slight macrocytic anemia with hematocrit 35% (normal range from 40% to 52%) and MCV 111 fL (normal range from 80 to 100 fL). White blood cell count is 4.3 × 103/mm^3 (normal range from 3.9 to 11.7 × 103/mm^3), and the differential reveals moderate elevation of eosinophils at 9% (normal range from 0% to 4%). Occasional basophilic stippling is noted on the peripheral smear. Liver transaminases are slightly elevated. BUN, creatinine, and urinalysis are normal.

14.4 Exposure Pathways

Environmental sources of arsenic exposure are food, water, and air. Commercial products like wood preservatives, pesticides, herbicides (weed killers, defoliants), fungicides, paints and pigments, leaded gasoline, fire salts (multicolored flame), food like wine (grapes sprayed with arsenic-containing pesticides), tobacco (plants sprayed with arsenic-containing pesticides), seafood (especially bivalves, certain cold water and bottom-feeding finfish, seaweed), industrial processes like purifying industrial gases (removal of sulfur), burning fossil fuels, and electronics manufacturing (microwave devices, lasers, light-emitting diodes, photoelectric cells, semiconductor devices), preserving animal hides. Workers in industries producing or using arsenic-containing compounds are potentially at risk.

Persons whose water supply contains high levels of arsenic or those living near sources of high ambient air levels of arsenic and workers in industries producing or using arsenic-containing compounds are potentially at risk and are at increased likelihood of exposure. Arsenic can cross the placenta, increasing the likelihood of exposure to the fetus, because it targets ubiquitous enzyme reactions, arsenic affects nearly all organ systems.

In animal models, the efficacy of chelation therapy generally declines as the time elapsed since exposure increases. If patients are treated within several hours after arsenic ingestion, chelation is likely to be beneficial. Therefore, even if arsenic ingestion is only suspected, it may be valuable to give one or two doses of dimercaprol (Dimercaprol (2,3-dimercaptopropanol, also known as British Anti-Lewisite or BAL) and (dimercaptosuccinic acid (DMSA) is the most frequently recommended chelating agent for arsenic) while awaiting confirmation since side effects usually are not noxious enough to outweigh benefits.

14.5 Conclusion

Environmental toxicology aids in addressing health disparities and contributes to public health by promoting awareness and education. By explaining the risks of exposure to toxic substances in accessible terms, the field empowers individuals to make informed decisions about their health. This education encourages advocacy for cleaner environments, safer products, and responsible industrial practices.

Environmental toxicology is a cornerstone of public health. It serves as an essential guardian of communities by identifying, assessing, and mitigating risks associated with environmental contaminants. In a world, where environmental challenges are increasingly complex, environmental toxicology plays an important role in protecting public health, promoting equitable outcomes, and fostering a cleaner and safer environment for all. Whether it is improving strength, corrosion resistance, ductility, or other crucial characteristics, alloying agents play a significant role in modifying the properties of materials to meet the ever-evolving demands of modern industries. As we face a growing number of environmental threats, this discipline remains at the forefront of the fight for healthier, more sustainable communities. Environmental toxicologists often collaborate with epidemiologists to conduct studies that link environmental exposures to health outcomes. This research helps identify associations between environmental factors and diseases, such as cancer, respiratory problems, and developmental issues.

Chapter 15

Carcinogenesis: Environmental Factors in Cancer Development

"From environmental hazards to cancer's front lines"

15.1 Introduction to Carcinogenesis

Carcinogenesis, often referred to as the process of cancer development, transforms normal cells into cancerous ones. At its core, carcinogenesis involves the disruption of the tightly regulated mechanisms that control cell growth, division, and differentiation. When these regulatory processes go awry, cells can acquire a series of genetic and epigenetic alterations that ultimately lead to uncontrolled proliferation and the formation of a malignant tumor (Fig. 15.1).

This process has been the subject of extensive research and exploration in the field of oncology. While genetic mutations play a central role, it is now widely recognized that environmental factors contribute significantly to cancer development.

15.2 Chemical Carcinogens

Chemical carcinogens are substances or compounds, both natural and synthetic, that have the potential to cause cancer. They exert

Environmental Toxicology: Concepts and Principles
Mohd Salim Reshi, Darakhshan Javaid, Syed Sanober Qadri, and Neelu Sinha

ISBN 978-981-5129-68-7 (Hardcover), 978-1-003-63707-3 (eBook)
www.jennystanford.com

their carcinogenic effects by interacting with cellular DNA, causing mutations or alterations in genes that regulate cell growth and division. These changes can lead to the uncontrolled proliferation of cells, ultimately resulting in the formation of tumors.

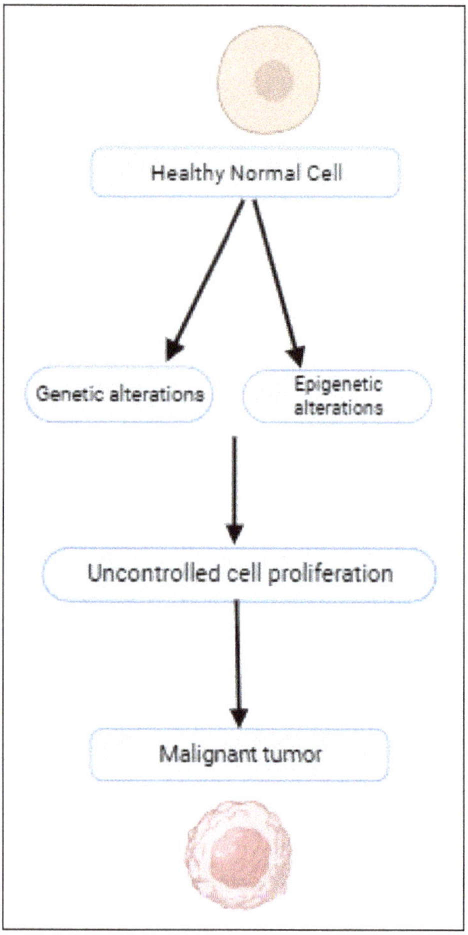

Figure 15.1 Formation of malignant tumor.

Chemical carcinogens can be found in various environments, including workplaces, homes, and the natural world. They are broadly classified into two main categories:

1. Genotoxic Carcinogens: These substances directly damage DNA, leading to genetic mutations. Examples include certain types of industrial chemicals, environmental pollutants, and some naturally occurring compounds like aflatoxins produced by certain molds.

2. Non-genotoxic Carcinogens: These substances do not directly damage DNA but promote cancer development through other mechanisms. They may influence cell signaling pathways, disrupt hormonal balances, or induce chronic inflammation. Examples include certain hormones, certain medications, and some naturally occurring substances like certain plant alkaloids.

15.3 Radiation and Carcinogenesis

Radiation, a ubiquitous force in our environment, has a profound impact on carcinogenesis. Both ionizing and non-ionizing forms of radiation can influence cellular processes, potentially leading to the initiation and progression of cancer. Ionizing radiation, emanating from sources like X-rays, nuclear materials, and certain radioactive isotopes, possesses sufficient energy to dislodge electrons from atoms, creating reactive ions that can directly damage DNA. This can result in genetic mutations, chromosomal aberrations, and disruptions in crucial cellular processes, ultimately contributing to the development of malignancies. Non-ionizing radiation, including ultraviolet (UV) rays from sunlight and radiofrequency electromagnetic fields, operates at lower energy levels and primarily induces cancer through indirect mechanisms, such as generating reactive oxygen species and promoting chronic inflammation. Some types of radiation are discussed below:

1. Ionizing Radiation: Ionizing radiation is a high-energy form of radiation that has the capability to remove tightly bound electrons from atoms, resulting in the formation of charged particles or ions. This property enables ionizing radiation to interact with cellular DNA, potentially leading to DNA damage and genetic mutations. Sources of ionizing radiation include medical procedures (such as X-rays and radiation therapy), nuclear power plants, and environmental sources like radon gas.

2. Ultraviolet (UV) Radiation: Ultraviolet radiation, emitted by the sun, is a form of non-ionizing radiation. Prolonged exposure to UV radiation can lead to DNA damage, particularly in the form of pyrimidine dimers, which can initiate carcinogenic processes. Skin cancer, including melanoma, is a well-documented outcome of excessive UV exposure.

3. Medical Radiation: While crucial for diagnostic and therapeutic purposes, medical radiation can inadvertently increase the risk of cancer. Diagnostic X-rays and computed tomography (CT) scans involve ionizing radiation, and repeated or unnecessary exposures should be carefully considered to minimize potential risks.

4. Occupational and Environmental Exposures: Certain occupations and environments may entail increased exposure to radiation. Workers in nuclear power plants, radiology departments, and certain industries may face higher risks. Additionally, individuals residing in regions with naturally occurring high levels of radiation, such as certain geological formations, may experience heightened exposure.

5. Genetic Susceptibility and Radiation: Individuals may vary in their susceptibility to the carcinogenic effects of radiation due to genetic factors. Some individuals possess genetic variations that render them more susceptible to radiation-induced DNA damage and subsequent cancer development.

6. Radiation Dose and Cancer Risk: The risk of radiation-induced cancer is influenced by factors such as the type of radiation, dose received, duration of exposure, and an individual's age. Understanding the dose-response relationship is crucial for accurately assessing cancer risks associated with different levels of radiation exposure.

Radiation, both ionizing and non-ionizing, stands as a significant environmental factor in carcinogenesis. Through various mechanisms, radiation can induce genetic alterations and cellular damage, ultimately contributing to cancer development. By comprehensively understanding the impact of radiation on carcinogenesis, we can implement strategies to minimize unnecessary exposure and develop more effective cancer prevention and management approaches.

15.4 Lifestyle and Behavioral Factors

Lifestyle and behavioral factors play a pivotal role in the development and progression of cancer. Choices related to diet, physical activity, tobacco and alcohol consumption, and other behaviors can significantly influence an individual's risk of developing various types of cancer. This section provides an in-depth exploration of the impact of lifestyle and behavioral factors on carcinogenesis.

1. Tobacco Use: Tobacco use, particularly smoking, is one of the most well-established and significant risk factors for cancer. It is linked to a multitude of cancer types, including lung, mouth, throat, esophagus, pancreas, bladder, and cervix. The carcinogenic compounds in tobacco smoke directly damage DNA and trigger genetic mutations that lead to cancer.

2. Diet and Nutrition: Dietary habits can both mitigate and exacerbate cancer risk. A diet rich in fruits, vegetables, whole grains, and lean proteins provides essential nutrients and antioxidants that support cellular health and reduce inflammation, thus lowering cancer risk. Conversely, diets high in processed foods, red and processed meats, and sugary beverages have been associated with an increased risk of certain cancers.

3. Physical Activity: Regular physical activity offers numerous health benefits, including reducing the risk of cancer. Exercise helps regulate hormones, improve immune function, and promote healthy body weight—all of which contribute to a lower risk of cancer development. Sedentary behavior, on the other hand, has been linked to an increased risk of several cancer types.

4. Alcohol Consumption: Excessive alcohol consumption is a significant contributor to cancer risk. It is associated with an increased likelihood of developing cancers of the mouth, throat, esophagus, liver, colon, and breast. Alcohol can directly damage DNA and lead to chronic inflammation and hormonal imbalances that promote carcinogenesis.

5. Sun Exposure and Skin Protection: Overexposure to ultraviolet (UV) radiation from sunlight is a major risk factor for skin cancer, including melanoma. Using protective measures such as sunscreen, wearing protective clothing, and avoiding peak sun hours can significantly reduce this risk.

6. Infectious Agents: Certain lifestyle behaviors, such as engaging in unprotected sexual activity or sharing needles for drug use, can lead to exposure to infectious agents like human papillomavirus (HPV) or hepatitis B and C, which are associated with increased cancer risk.

7. Screening and Early Detection: Engaging in regular cancer screenings, such as mammograms and colonoscopies, is a positive behavior that can lead to the early detection and treatment of cancer, significantly improving outcomes.

8. Stress and Coping Mechanisms: Chronic stress and ineffective coping mechanisms can have negative effects on the immune system and promote inflammation, potentially contributing to cancer development. Adopting healthy stress management techniques, such as mindfulness, exercise, and social support, can mitigate these risks.

Understanding the profound influence of lifestyle and behavioral factors on carcinogenesis empowers individuals to make informed choices that can significantly reduce their risk of developing cancer. Public health initiatives aimed at promoting healthy behaviors and providing education on cancer prevention are essential components of comprehensive cancer control efforts.

15.5 Infectious Agents and Cancer

Infectious agents, including viruses, bacteria, and parasites, have emerged as significant players in the complex landscape of carcinogenesis. These pathogens possess the capacity to introduce genetic alterations, promote chronic inflammation, and disrupt cellular regulatory mechanisms, all of which can contribute to the initiation and progression of cancer. For instance, certain strains of human papillomavirus (HPV) have been closely linked to cervical and oropharyngeal cancers, while hepatitis B and C viruses are notorious culprits in liver cancer development. *Helicobacter pylori*, a bacterium residing in the stomach lining, is associated with an increased risk of gastric cancer. Additionally, Schistosoma parasites have been implicated in the development of bladder cancer.

1. Human Papillomavirus (HPV): HPV is a group of related viruses that can cause various types of cancers, including cervical, anal, and oropharyngeal cancers. High-risk strains of HPV, such as HPV 16 and 18, are known to integrate their DNA into the host cell's genome, leading to aberrant cellular growth and division.

2. Hepatitis B and C Viruses (HBV and HCV): Chronic infection with HBV or HCV is a major risk factor for liver cancer. These viruses can cause prolonged liver inflammation, leading to the development of cirrhosis and eventually hepatocellular carcinoma.

3. Epstein-Barr Virus (EBV): EBV is associated with several types of cancer, including Burkitt lymphoma, nasopharyngeal carcinoma, and Hodgkin lymphoma. The virus can integrate its genetic material into host cells, disrupting normal cellular processes.

4. Human Immunodeficiency Virus (HIV): HIV weakens the immune system, making individuals more susceptible to various infections, including those caused by cancer-causing agents. This increased susceptibility can lead to a higher incidence of certain cancers, such as Kaposi sarcoma, non-Hodgkin lymphoma, and cervical cancer.

5. *Helicobacter pylori*: This bacterium is a known risk factor for gastric cancer. Chronic infection with *H. pylori* leads to persistent inflammation in the stomach lining, which can ultimately lead to the development of cancerous lesions.

6. *Schistosoma haematobium*: This parasitic worm is linked to an increased risk of bladder cancer. Chronic infection with the parasite leads to chronic inflammation of the bladder lining, which can result in the development of cancer.

7. Human T-lymphotropic Virus Type 1 (HTLV-1): HTLV-1 is associated with adult T-cell leukemia/lymphoma, a rare form of cancer of the immune system's T cells. The virus directly affects T-cell function, leading to uncontrolled cellular growth.

8. Merkel Cell Polyomavirus (MCV): MCV is a recently discovered virus that is associated with the development of Merkel cell carcinoma, a rare and aggressive skin cancer. The virus integrates its genetic material into host cells, contributing to cancer formation. Vaccination against viruses like HPV and HBV, early detection and treatment of infections, and targeted therapies for virus-associated cancers are all important strategies in managing the impact of infectious agents on carcinogenesis. Table 15.1 shows the infectious agents and the cancers that are caused by them.

Table 15.1 The infectious agents and types of cancers caused by them

S. No.	Infectious Agents	Types of Cancer
1.	Human Papillomavirus (HPV)	Cervical, anal, and oropharyngeal cancers
2.	Hepatitis B and C Viruses	Liver cancer
3.	Epstein-Barr Virus (EBV)	Burkitt lymphoma, nasopharyngeal carcinoma, and Hodgkin lymphoma
4.	Human Immunodeficiency Virus (HIV)	Kaposi sarcoma, non-Hodgkin lymphoma, and cervical cancer
5.	*Helicobacter pylori*	Gastric cancer
6.	*Schistosoma haematobium*	Bladder cancer
7.	Human T-lymphotropic Virus Type 1 (HTLV-1)	Adult T-cell leukemia/ lymphoma
8.	Merkel Cell Polyomavirus (MCV)	Merkel cell carcinoma (skin cancer)

15.6 Hormonal Influences on Carcinogenesis

Hormones play a pivotal role in regulating various physiological processes within the body, including cell growth and differentiation. Consequently, hormonal imbalances or exposure to exogenous hormones can significantly influence the development of cancer. Estrogen, for example, has been linked to an increased risk of breast and endometrial cancers, as it can stimulate the growth of hormone-sensitive cells in these tissues. Similarly, androgens, a group of male sex hormones, have been associated with the development of prostate cancer. Hormonal therapies, such as those used in menopausal hormone replacement or for treating conditions like polycystic ovary syndrome, can also impact cancer risk.

1. Estrogen and Progesterone: Estrogen and progesterone are sex hormones primarily produced by the ovaries in females, but they are also present in smaller amounts in males. Prolonged exposure to high levels of estrogen, often seen in conditions like early menarche, late menopause, and hormone replacement therapy, is associated with an increased risk of breast and endometrial cancers (Fig. 15.2). These hormones can stimulate cell proliferation in target tissues, potentially leading to the accumulation of genetic mutations and cancer development.

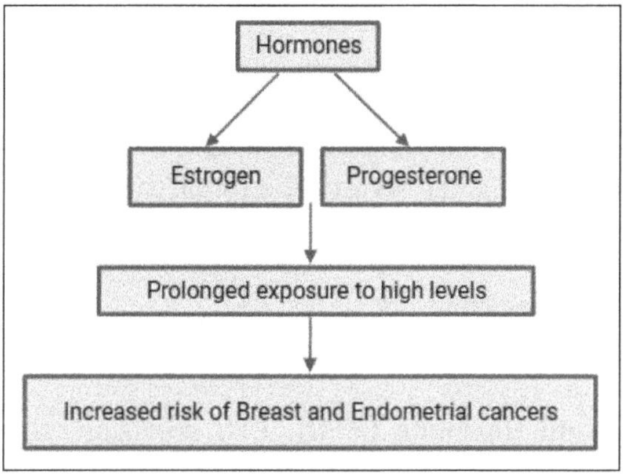

Figure 15.2 Prolonged exposure to estrogen and progesterone.

2. Androgens: Androgens, predominantly testosterone, are male sex hormones produced by the testes and in smaller quantities by the adrenal glands in both males and females. Elevated levels of androgens are linked to an increased risk of prostate cancer. Androgens can stimulate the growth of prostate cells, and in the presence of other genetic or environmental factors, this increased cell division can contribute to the development of cancer.

Androgens → Elevated levels linked to prostate cancer

3. Thyroid Hormones: Thyroid hormones play a crucial role in regulating metabolism. Disorders related to thyroid function, such as hyperthyroidism (excess thyroid hormone) or hypothyroidism (insufficient thyroid hormone), can influence the risk of thyroid cancer. Additionally, exposure to high levels of ionizing radiation, which affects the thyroid gland, can increase the risk of thyroid cancer.

4. Insulin and Insulin-like Growth Factor (IGF): Insulin, a hormone produced by the pancreas, regulates blood glucose levels. Elevated levels of insulin, often associated with conditions like obesity and insulin resistance, can promote cell proliferation and inhibit cell death, potentially contributing to the development of various cancers, including breast, colon, and pancreatic cancers.

Insulin & IGF ⟶ Elevated levels linked to various cancers (breast, colon, pancreatic cancers)

5. Prolactin: Prolactin is a hormone primarily known for its role in lactation. Elevated levels of prolactin, often seen in conditions like hyperprolactinemia, have been associated with an increased risk of certain cancers, including breast and ovarian cancers.

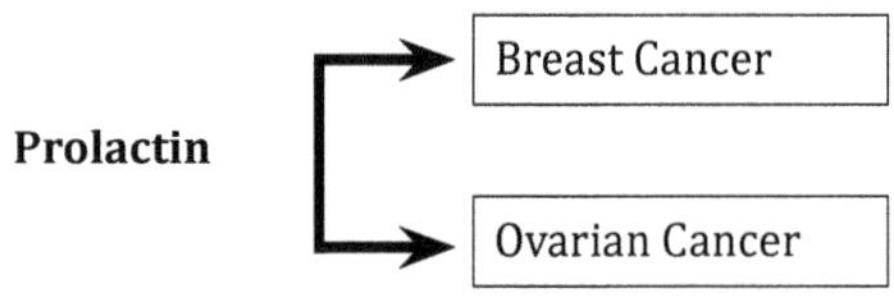

6. Parathyroid Hormone (PTH): PTH regulates calcium levels in the blood. Hyperparathyroidism, a condition characterized by excessive PTH secretion, can lead to an increased risk of certain cancers, particularly those of the kidney.

Understanding the influence of hormonal factors on carcinogenesis is essential for identifying individuals at higher risk for certain cancers. Hormone-related cancers can be targeted through preventive measures, such as hormone replacement therapy management, lifestyle modifications, and pharmacological interventions. Additionally, ongoing research in this field holds promise for the development of novel approaches to hormone-related cancer prevention and treatment.

Parathyroid Hormone ⟶ Kidney cancer

15.7 Epigenetic Modifications and Environmental Exposures

Epigenetic modifications refer to reversible alterations in gene expression patterns that do not involve changes in the underlying DNA sequence. These modifications play a pivotal role in how environmental exposures can influence an individual's susceptibility to various diseases, including cancer. Environmental factors, such as chemical pollutants, dietary components, and lifestyle choices, can impact epigenetic processes. DNA methylation, histone modifications, and non-coding RNA molecules are key players in this epigenetic regulation. For example, exposure to carcinogenic compounds like tobacco smoke has been linked to altered DNA methylation patterns, which can lead to the dysregulation of genes involved in cancer development. Similarly, dietary components like folate and certain phytochemicals have demonstrated the ability to modulate epigenetic marks, potentially affecting cancer risk. The relationship between epigenetic modifications and environmental exposures provides crucial insights into the molecular mechanisms underlying carcinogenesis. This knowledge holds significant promise for the development of targeted interventions and personalized therapeutic strategies to mitigate the impact of environmental exposures on cancer susceptibility.

1. DNA Methylation: DNA methylation involves the addition of a methyl group to a cytosine base in DNA, typically occurring in regions known as CpG islands. Hypermethylation of promoter regions can lead to gene silencing, inhibiting the expression of tumor suppressor genes. Environmental exposures, such as exposure to tobacco smoke or certain chemicals, can lead to aberrant DNA methylation patterns, contributing to cancer initiation and progression.

2. Histone Modifications: Histone proteins play a role in packaging DNA within the cell nucleus. Chemical modifications to histones, such as acetylation, methylation, and phosphorylation, can alter chromatin structure and accessibility, influencing gene expression. Environmental factors, including exposure to pollutants or dietary components, can lead to changes in histone modifications, potentially impacting cancer-related genes.

3. Non-Coding RNAs: Non-coding RNAs, including microRNAs (miRNAs) and long non-coding RNAs (lncRNAs), are involved in post-transcriptional gene regulation. They can influence cancer-related pathways by targeting specific mRNA molecules. Environmental exposures, such as radiation or chemical toxins, can lead to dysregulation of non-coding RNA expression, potentially contributing to cancer development.

4. Chromatin Remodeling: Chromatin remodeling complexes modify the structure of chromatin, allowing for access to DNA for transcription. Environmental factors, including exposure to carcinogens or hormonal imbalances, can lead to alterations in chromatin remodeling patterns, affecting gene expression profiles related to cancer.

5. Environmental Exposures and Epigenetic Alterations: Various environmental exposures have been linked to specific epigenetic changes associated with cancer. For example, exposure to heavy metals, pesticides, air pollutants, and certain dietary components can lead to epigenetic modifications that influence cancer-related pathways.

6. Developmental Exposures: Environmental exposures during critical periods of development, such as in utero or during early childhood, can have lasting effects on epigenetic patterns. These alterations may increase the susceptibility to cancer later in life.

7. Lifestyle Factors: Lifestyle choices, including diet, physical activity, and exposure to stress, can influence epigenetic modifications. For example, a diet rich in certain nutrients and phytochemicals can impact DNA methylation patterns. Environmental exposures and epigenetic modifications help us understand the mechanisms by which external factors contribute to cancer development. This knowledge is essential for the development of targeted interventions aimed at mitigating epigenetic alterations associated with carcinogenesis.

15.8 Gene-Environment Interactions

Gene-environment interactions refer to the relationship between an individual's genetic makeup and external environmental factors in influencing an individual's susceptibility to disease, including cancer. While genetic predispositions can play a significant role in disease risk, the expression and impact of these genes are often modulated by environmental exposures. These interactions can either amplify or mitigate the effects of genetic variations on disease development. For instance, certain genetic mutations may confer an increased risk of cancer, but this risk may be heightened or attenuated by specific environmental factors. Conversely, some individuals with a genetic predisposition may never develop the disease if they are not exposed to relevant environmental triggers.

15.8.1 Gene-Environment Interactions in Carcinogenesis

Gene-environment interactions refer to the combined effects of genetic variations and external factors in influencing an individual's susceptibility to cancer. The genetic factors, environmental exposures, and their joint impact on carcinogenesis are discussed here.

1. **Genetic Variations:** Genetic variations, including single nucleotide polymorphisms (SNPs) and mutations, can influence an individual's susceptibility to cancer. These variations can alter the function of key genes involved in cell growth, DNA repair, and other cellular processes. Some genetic variations may confer an increased risk of cancer, while others may provide protective effects.

2. **Environmental Exposures:** Environmental factors encompass a wide range of exposures, including lifestyle choices, occupational hazards, chemical pollutants, radiation, and infectious agents. These exposures can directly damage DNA, disrupt cellular processes, or create conditions conducive to carcinogenesis.

3. **Susceptibility Genes:** Certain genes, known as susceptibility genes, may increase an individual's vulnerability to specific environmental exposures. For example, individuals with specific variants in genes involved in DNA repair pathways may be more susceptible to the carcinogenic effects of UV radiation from sunlight.

4. **Polygenic Risk Scores:** Polygenic risk scores aggregate information from multiple genetic variants associated with cancer risk. These scores provide a more comprehensive assessment of an individual's genetic predisposition to cancer and can be combined with information on environmental exposures for a more accurate risk assessment.

5. **Gene-Environment Interactions in Specific Cancers:**

- **Lung Cancer:** The interaction between genetic variations, particularly in genes related to tobacco metabolism and detoxification, and cigarette smoke exposure significantly influences lung cancer risk.

- **Breast Cancer:** Genetic variations in genes such as BRCA1 and BRCA2, combined with factors like hormonal exposure, reproductive history, and lifestyle choices, play a crucial role in breast cancer susceptibility.
- **Colorectal Cancer:** Interactions between genetic variations in genes like APC and environmental factors like diet, particularly consumption of red and processed meats, influence the risk of colorectal cancer.
- **Skin Cancer:** Genetic variations in genes related to DNA repair, combined with UV radiation exposure, are critical factors in the development of skin cancer.

6. Epigenetic Modifications as Mediators: Epigenetic alterations can act as mediators in gene-environment interactions. Environmental exposures can induce epigenetic changes that modify gene expression, potentially influencing cancer risk. Gene-environment interactions are essential for personalized cancer risk assessment and prevention strategies. It allows for the identification of individuals who may be particularly vulnerable to specific environmental exposures. This knowledge also informs the development of targeted interventions and screening protocols tailored to an individual's genetic and environmental profile, ultimately improving cancer prevention and early detection efforts.

15.9 Environmental Carcinogenesis and Public Health

As our understanding of the impact of pollutants, radiation, lifestyle choices, and infectious agents on carcinogenesis deepens, it becomes increasingly crucial for public health initiatives to address these risks. Implementing stringent regulatory measures, promoting clean and sustainable practices, and raising awareness about modifiable risk factors are essential steps in mitigating environmental contributions to cancer incidence. Additionally, targeted interventions, such as vaccination programs for infection-related cancers and smoking cessation efforts, play a critical role in preventing environmentally-induced malignancies. By prioritizing research, education, and policy measures aimed at reducing environmental carcinogenic exposures, we strive to safeguard the health and well-being of communities worldwide. This comprehensive approach not only addresses the immediate health concerns but also fosters a sustainable and cancer-resilient environment for generations to come.

In conclusion, the study of carcinogenesis reveals a process influenced by a myriad of factors. Genetic mutations, environmental exposures, hormonal influences, epigenetic modifications, and gene-environment interactions all contribute to the complex process from normal cells to cancerous entities. Understanding these mechanisms not only provides us with knowledge about the origins of cancer but also paves the way for targeted prevention, early detection, and effective treatment strategies. Recognizing the pivotal role of environmental factors, including chemical carcinogens, radiation, infectious agents, and lifestyle choices, emphasizes the need for comprehensive public health initiatives. By implementing regulatory measures, advocating for clean and sustainable practices, and prioritizing education and awareness, we can significantly reduce environmental contributions to cancer incidence. Additionally, the integration of genetic and environmental risk assessments allows for tailored interventions and personalized care, further enhancing our ability to combat this formidable disease. Through collaborative efforts across the scientific, medical, and public health communities, we strive for a future where cancer prevention and treatment are not only more effective but also more precisely targeted, ultimately improving outcomes and quality of life for individuals affected by this global health challenge.

Chapter 16

Toward A Sustainable Future: Mitigating Toxicological Risks

"Mitigating toxic hazards for a healthier tomorrow"

16.1 Introduction

The aim of conducting a risk assessment is to gauge the potential impact of exposure to chemicals in the environment on both human health and the ecosystem. The Environmental Protection Agency (EPA) employs a four-step process in risk assessment, which encompasses:

1. Hazard assessment
2. Dose–response assessment
3. Exposure assessment
4. Risk characterization

16.1.1 Hazard Assessment

The current utilization of chemicals is immense. Hodgson and Guthrie's 1980 data revealed 1500 active pesticide ingredients, 4000 active therapeutic drug ingredients, 2000 drug additives, 2500 nutritional food additives, 3000 food additives for product longevity,

Environmental Toxicology: Concepts and Principles
Mohd Salim Reshi, Darakhshan Javaid, Syed Sanober Qadri, and Neelu Sinha

ISBN 978-981-5129-68-7 (Hardcover), 978-1-003-63707-3 (eBook)
www.jennystanford.com

and 50,000 common-use chemicals. Given the expansion of the chemical and pharmaceutical industries, these figures are likely much higher now. Past instances indicate that certain chemicals, though not harmful in large quantities, can exhibit mutagenic, carcinogenic, or adverse effects on reproductive and immune systems with chronic exposure to minute doses. To safeguard human health, it is crucial to ascertain that compounds encountered in daily life (like cosmetics, foods, and pesticides) do not pose long-term harm through regular or periodic exposure.

This chapter will primarily delve into the topics of carcinogenicity and mutagenicity, with a secondary consideration for endocrine disruptors. While some chemical carcinogenicity has been established through epidemiological studies, these studies face challenges due to the extended latency period of cancer, necessitating years before conclusive results are obtained. Moreover, they are resource-intensive.

An alternative method involves bioassays conducted on animals. Although effective in predicting human cancer hazards, these bioassays can take two years or more and require a minimum of 600 animals per assay. Given the considerable time and financial investment, this approach is impractical for large-scale screening. Therefore, there is a need for an economical and short-term assay system for the initial assessment of potential mutagens and carcinogens.

16.1.1.1 Bacterial mutagenesis test

Numerous versions of bacterial mutagenesis tests are available, with the Ames test being the most widely used. This test employs specially engineered strains of *Salmonella typhimurium* that cannot synthesize the amino acid histidine and, consequently, rely on histidine for growth. The test assesses the frequency of back mutations to a histidine-independent parent strain.

In this procedure, bacteria are cultured on agar plates containing minimal growth medium with a limited amount of histidine to allow background growth. The compound to be tested is also introduced. Back-mutated organisms form colonies that are then counted. Control plates are included to account for spontaneous mutations. A dose–response curve is generated by varying doses of the mutagen. Since many potential mutagenic-carcinogenic substances require metabolic activation, and bacteria lack such an activating system, liver microsomal preparation (post mitochondrial supernatant, PMS) is added to the plates.

Various mutated strains with distinct genetic makeup have been developed. These strains differentiate between base substitution and frameshift mutations. Additionally, supersensitive strains, lacking a DNA repair system or lipopolysaccharide coating, are more susceptible to external chemicals.

Experimental testing of the predictive reliability of the Ames assay has shown that 85% of known carcinogens tested positive, while compounds classified as non-carcinogens yielded positive results in fewer than 10% of cases. A more recent study indicated that the predictability of the Salmonella test varies significantly based on the chemical class of compounds tested. For example, only 40% of chlorinated carcinogens were identified as mutagens, while 75% and 100% of carcinogenic amines and nitro compounds, respectively, tested positive as mutagens.

Another bacterial assay involves *Escherichia coli* with a deficiency in a DNA repair mechanism. Mutagens that induce DNA lesions exhibit greater toxicity to the genetically altered strain than to the parent strain.

16.1.1.2 DNA repair assay

This assay, performed in mammalian cell culture, is designed to detect compounds injurious to DNA. The test presupposes that the injury to DNA stimulates the repair mechanism. DNA repair is measured by the increase in the incorporation of 3 H-thymidine into DNA above that of the control. The radioactivity is determined either by scintillation counting or by autoradiography. PMS is added to the cultures to activate precarcinogens. A modification of this procedure, the hepatocyte primary culture–DNA repair assay, uses freshly isolated, non-dividing liver cells. This system does not need PMS, as the hepatocytes can activate precarcinogens. In addition, the non-dividing cells have no background of thymidine incorporation.

16.1.1.3 Mammalian mutagenicity assays

There are three types of mammalian mutagenicity assays, with the first and most common one employing mammalian fibroblasts. In this assay, mutants are identified by the emergence of colonies resistant to purine analogs, specifically 6-thioguanine or 8-azaguanine (Fig. 16.1).

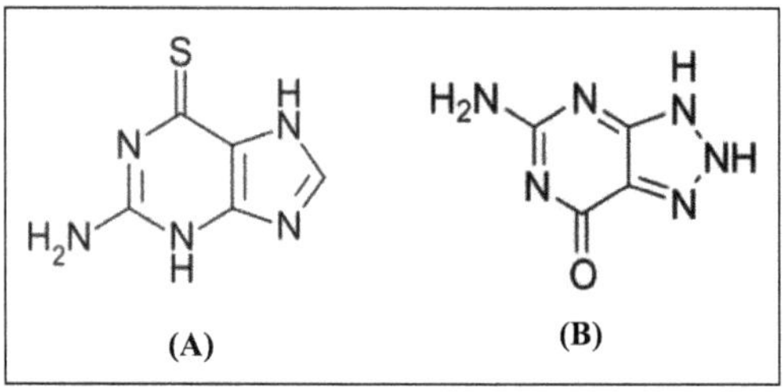

Figure 16.1 6-Thioguanine **(A)** and 8-azaguanine **(B)**.

These analogs are not inherently toxic but become cytotoxic when activated to nucleotides by the "purine salvage" enzyme, hypoxanthine–guanine phosphoribosyltransferase (HGPRT), present in most cells. HGPRT recycles preformed purines for nucleic acid synthesis, although it is not vital for cell survival since most cells can synthesize purines de novo. The mechanism of the HGPRT-catalyzed reaction is depicted in (Fig. 16.2).

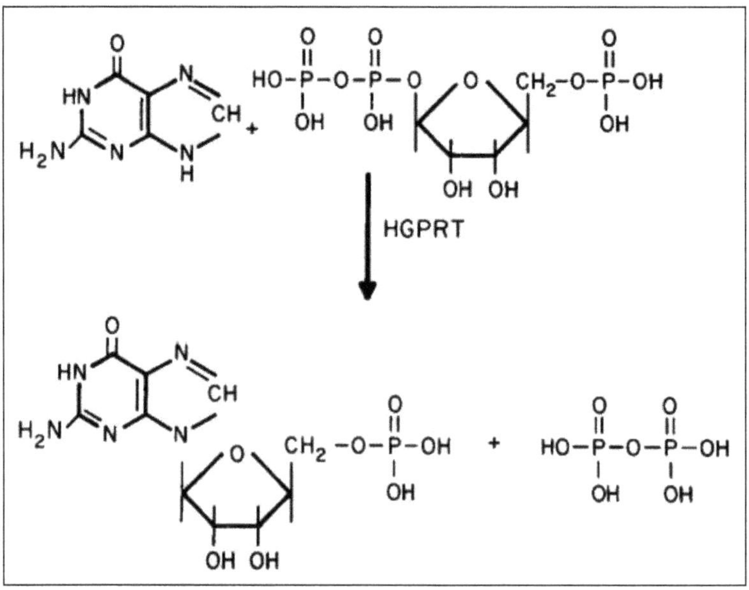

Figure 16.2 Salvage pathway of purines, a mechanism of activation of purine analogs.

Under normal conditions, cells do not grow in cultures exposed to either 6-thioguanine or 8-azaguanine. However, in the presence of a mutagen, mutants resistant to 6-thioguanine-8-azaguanine, lacking HGPRT, emerge, and colonies form. The addition of post-mitochondrial supernatant (PMS) is necessary to activate pre-carcinogens. This assay is highly sensitive because HGPRT is not an essential enzyme, and its deletion does not lead to the formation of lethal mutants. The HGPRT locus is on the X (sex) chromosome, which is mutable and lacks duplication. A modified version of this procedure incorporates freshly prepared hepatocytes as a feeder layer, omitting post-mitochondrial supernatant (PMS) since hepatocytes possess xenobiotic-activating enzymes.

Another mammalian mutagenesis assay focuses on mutations in the locus responsible for thymidine kinase synthesis, an enzyme necessary for the activation of the antimetabolite iododeoxyuridine (Fig. 16.3). Only mutated cells, lacking kinase, form colonies in the presence of the antimetabolite. As kinase is not an essential enzyme, no lethal mutants are generated.

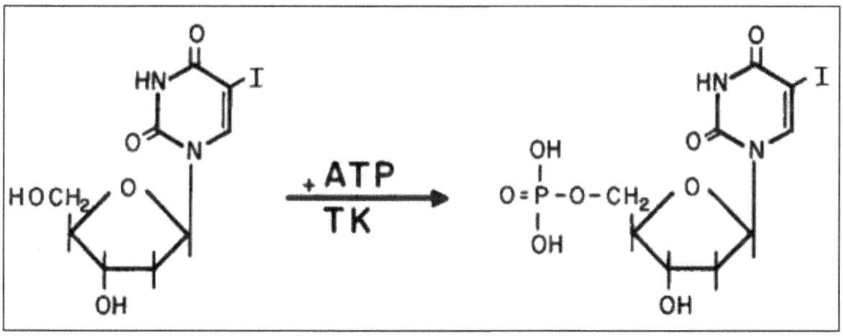

Figure 16.3 Mechanism of activation of 5-iododeoxyuridine.

The third assay in this category identifies mutants resistant to the alkaloid ouabain. Ouabain resistance results from a mutation in the gene responsible for membrane ATP-ase synthesis, an enzyme involved in K^+-Na^+ transfer. Ouabain non-competitively inhibits this enzyme, disrupting essential cell functions. However, this assay lacks sensitivity because only mutants with ATP-ase altered to retain enzymatic activity but not bind ouabain are available for scoring. Mutants with inactive ATP-ase are lethal and do not form colonies.

The discussed assays focus on mutagens and, by implication, genotoxic carcinogens. The subsequent two assay systems are applicable to both genotoxic and epigenetic carcinogens.

16.1.1.4 Sister chromatid exchange assay

This assay assesses the exchange of loci between sister chromatids of chromosomes. Cells are cultured in the presence of 5-bromodeoxyuridine (5BrdUR) for a period encompassing two rounds of DNA replication. 5BrdUR is integrated into the newly synthesized DNA strand instead of thymidine. Following the first replication, one DNA strand of one chromatid contains 5BrdUR; after the second replication, both strands of one chromatid and one strand of the second chromatid contain 5BrdUR. Using fluorescent staining techniques, the two chromatids can be differentiated. This procedure enables the observation of mutagen-induced exchanges of chromatid segments. While highly sensitive, the specific chemical injuries responsible for these chromosomal lesions are often unidentified in most cases.

16.1.1.5 Cell transformation assay

The cell transformation assay stands out as the only test directly evaluating malignant transformation rather than mutagenesis, unlike the assays discussed thus far. It is applicable to both genotoxic and epigenetic carcinogens.

In this assay, mammalian cells are cultivated on agar as a monolayer. Upon reaching confluence, the growth of normal cells is halted by contact inhibition. However, in the presence of a carcinogen, cells undergoing malignant transformation continue to divide. Since there is limited space for proliferation in the horizontal plane, the transformed cells stack on top of each other, making colonies easily quantifiable. Post mitochondrial supernatant (PMS) is required to activate pre-carcinogens in the culture. Injecting these proliferating cells into animals leads to tumor formation, providing evidence that the colonies indeed represent malignantly transformed cells.

16.1.1.6 Carcinogenicity testing in fish

Several test systems for carcinogens use fish. With these systems, there is no need for elaborate cage sterilization and bedding changes. Thus, a much larger number of animals can be used at a lower cost. A comprehensive review of this subject has been published.

16.1.1.7 Biological testing in rodents

Bioassays, involving the testing of chemicals in laboratory animals, provide reliable information about carcinogenicity. Despite species differences in susceptibility to carcinogens, it is possible to replicate every human cancer in animals, as most animals are susceptible to cancer. However, due to the substantial cost of bioassays (estimated by the EPA to range from $390,000 to $980,000 per assay) and the lengthy time involved (up to 30 months), it is not practical to test all 50,000 compounds in common use. Consequently, a selection process for bioassay testing of chemicals has been implemented.

Currently, two testing programs operate in the United States. The Interagency Testing Committee, consisting of eight members from different federal agencies and departments, recommends chemicals to the EPA Administrator for testing. The National Toxicology Program (NTP) Chemical Nomination and Selection Committee reports to the National Cancer Institute (NCI).

The selection for bioassay is based on the outcomes of multiple in vitro tests and consideration of chemical structures. Structure–activity relationship (SAR) studies have been conducted with various compound classes, and reasonably accurate predictions regarding the potential carcinogenicity of a compound can be made, at least within some groups.

The NCI has issued Guidelines for Carcinogenic Bioassay in Small Rodents, outlining the minimum requirements for the design and execution of a bioassay. The essence of these guidelines, in condensed form, is provided.

- Each chemical should be tested in at least two species and both sexes (rats and mice are usually used).
- Each bioassay should contain at least 50 animals in each experimental group.
- Exposure to chemicals should start when the animals are 6 weeks old (or younger) and continue for most of their life span (for mice and rats, usually 24 months). The observation period should continue for 3–6 months after administration of the last dose.
- One treatment group should receive the maximum tolerated dose (MTD), which is defined as the highest dose that can be

given that would not alter the animals' normal life span from effects other than cancer. The other treatment group is treated with a fraction of the MTD.

- The route by which a chemical is administered should be the same or as close as possible to the one by which human exposure occurs. The chemicals may be given by any of the following routes: orally (with food or water or by force-feeding), by inhalation, or topically (by application to the skin).

In some cases, two-generation bioassays are conducted, exposing both generations to a potential carcinogen. The advantage of this approach lies in its exposure to fetuses and very young animals, which are more sensitive to chemical injury than adults. Animals that die during the study and survivors euthanized at its conclusion are examined for tumors. The results are statistically evaluated with a p-value of 0.05, signifying that the probability of obtaining the given results by chance is less than 5%.

A positive outcome in a bioassay indicates but does not necessarily prove, that an agent will be carcinogenic in humans. As of 1982, the International Agency for Research on Cancer (IARC) listed 142 substances experimentally proven to be carcinogens in animals. However, only 14 of these have been officially recognized as human carcinogens.

16.1.2 Dose–Response Assessment

When extrapolating from bioassay-generated dose–response data to obtain a quantitative estimate of human risk, two parameters must be considered: biological extrapolation and numerical extrapolation.

16.1.2.1 Biological extrapolation

Metabolic differences between humans and test animals, along with the usual high level of inbreeding in laboratory animals compared to the genetic heterogeneity in the human population, create a fundamental challenge in adjusting the dose measured in bioassays to the dose experienced by humans. Several approaches can be considered to address this issue:

- Straight translation from animals to humans of milligrams per kilogram per day.
- Straight translation from animals to humans of milligrams per square meter per day.
- Straight translation from animals to humans of milligrams per kilogram per lifetime.
- In cases where the experimental dose is measured as parts per million (ppm) in food, water, or air, human exposure is expressed in the same units.

A study conducted by the National Research Council (NRC) compared the incidence of site-specific chemical-induced tumors in experimental animals and humans, evaluating five chemicals and cigarette smoking. The translation from bioassay to humans was based on milligrams per kilogram per lifetime. For two chemicals (N, N-bis(2-chloroethyl)-2-naphthylamine and benzidine) and cigarette smoking, the human incidence aligned with predictions from the animal study. However, for aflatoxin B1, diethylstilbestrol, and vinyl chloride, the human incidence was significantly overestimated (10, 50, and 500 times, respectively). As a result, the Consultative Panel on Health Hazards of Chemicals and Pesticides concluded: "While there are major uncertainties in extrapolating the results of animal tests to man, this is usually the only available method. Despite the uncertainties, enough is known to indicate what dependencies on dose and time may operate and to provide rough predictions of induced cancer rates in the human population."

Another challenge highlighted is the interpretation of bioassay results when the response of the two tested animal species varies greatly or if only one species responds positively. Despite some controversy about how to handle such data, there is general agreement among U.S. federal agencies that the extrapolation should be based on results from the more sensitive species.

16.1.2.2 Numeric extrapolation

In bioassays, meaningful results often require exposing test animals to relatively high doses of potential carcinogens. A typical dose–response relationship can be demonstrated for high doses. However, humans are usually exposed to much lower doses of environmental carcinogens than those used in laboratory animals. The cancer incidence resulting from such low exposure is expected to be significantly lower than observed in bioassays. The trajectory of the dose–response curve for exposures below the lowest observable bioassay exposure is uncertain and can only be speculated upon. Due to this uncertainty, the most used extrapolation method is a straight line from the lowest observed dose–effect point to the zero dose. However, alternative approaches have been proposed.

The infra-linear extrapolation (depicted in Fig. 16.4) can be derived from models based on fitting observable data points into a mathematical equation. Unfortunately, in practice, there are typically very few observable data points available, often just two or three. This scarcity means that many models can equally well fit the experimental dose–response curve, rendering the extrapolated segment highly hypothetical. The challenge lies in accurately predicting the dose–response relationship at lower exposure levels where direct observations are limited.

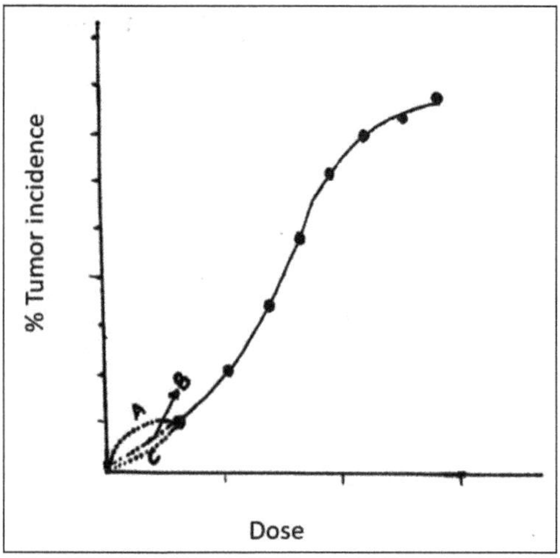

Figure 16.4 Possible ways of extrapolation of a bioassay dose–response curve. Key: A – super-linear, B – linear, and C – infra-linear.

Super-linear extrapolation can generate a series of hypothetical lines without providing a logical basis to place greater confidence in any specific line. The super-linear model concept arises from the observation that, in some bioassays, lower doses were more effective in producing tumors than higher doses. However, a critical issue with this approach is that the reduced effectiveness of higher doses is often attributed to the agent's toxicity, leading to a higher mortality rate among animals before tumor development.

Dose–response assessment significantly varies depending on the extrapolation model used for the dose–response curve. Compared to straight-line extrapolation, the infra-linear model tends to underestimate, while the super-linear model tends to overestimate tumor incidence.

All these models operate on the generally accepted notion that there is no threshold dose below which tumors do not occur. This point cannot be definitively proven or disproven; even if tumors cannot be demonstrated below a certain dose, it is conceivable that with a larger sample size, some tumors might appear. The lack of a clear threshold adds complexity to the assessment of the relationship between dose and tumor incidence.

16.1.2.3 Negative results

The fact that tumors were not detected in a test population of 100 animals does not indicate zero risk. According to statistical calculations, the absence of tumors indicates merely that there is a 95% likelihood that the actual incidence of tumors is no more than 0.45%. This estimate of tumor incidence that might escape detection represents the upper 95% confidence limit of an experiment with 100 animals. It must be concluded that quantitative cancer risk assessment of environmental toxins is highly hypothetical.

16.1.3 Exposure Assessment

The factors to be considered in exposure assessment are:

1. Who and what is likely to be exposed to the compound?
2. How much exposure may be anticipated?
3. In which way, how long, and under what circumstances will the exposure occur?
4. To make calculations of the overall human exposure possible, certain standard values of human anatomy and physiology are set:
    - Mass (kg) – man: 70, woman: 60, child: 20.
    - Skin surface area (m^2) – total (180 cm tall): 1.8, clothed with short sleeves: 0.3, clothed with long sleeves: 0.1.

- Resting respiration rate (L/min) – man: 7.5, woman: 6, child: 4.8.
- Respiration rate during light activity (L/min) – man: 20, woman: 19, child: 13.
- Volume of air breathed (m^3/day) – man: 23, woman: 21, child: 15.
- Fluid consumption (L/day) – man: 2, woman: 1.4, child: 1.4.
- Food consumption (g/day) – all humans: 1,500.

Exposure assessment is a challenging task, primarily due to the mobility of individuals, engagement in diverse activities, and the need to consider all potential routes of exposure. For example, to estimate the overall human exposure to a carcinogen from contaminated groundwater, calculations must account for contributions from the following routes of exposure:

- Direct exposure through drinking.
- Exposure through inhalation from showering, bathing, and other uses of water.
- Exposure through the skin by the body's contact with the contaminated water.
- Exposure through ingestion of food that was in contact with the contaminated water.

Moreover, at each stage of this analysis, the bioavailability for each route of entry and the metabolism of the carcinogen must be considered. A general criticism of the exposure assessment is that it is done for each carcinogen separately; whereas, in a real-life situation, people may be exposed to several carcinogens at the same time. A cumulative exposure may have an additive, synergistic, or antagonistic effect. In addition, simultaneous exposure to inhibitors or inducers of xenobiotic-metabolizing enzymes may complicate the true picture even further.

16.1.4 Risk Characterization

Cancer risk can be expressed in various ways, and the most common measure is the individual lifetime risk. This conveys the probability, such as 1 in 10,000 or 1 in 100,000, that an individual will develop cancer during their lifetime due to continuous exposure to a carcinogen. Using the straight-line extrapolation of the dose–response curve to zero, the percentage of cancers per unit of the carcinogen is calculated, known as carcinogenic "potency" or "unit cancer risk." Multiplying this potency by the exposure dose yields the individual lifetime risk.

Population or societal risk is derived by multiplying the individual risk by the number of people exposed. It represents the number of cases attributed to one-year or lifetime exposure to a carcinogen. Defining the time parameter is crucial, as the results vary significantly depending on whether the calculation is for a year or a lifetime.

Relative risk is expressed by dividing the risk (incidence rate) in the exposed group by the risk in the unexposed group or the general population.

The risk in the exposed group divided by the risk in the general population, adjusted for factors like age and time period, is termed the "standardized mortality (or morbidity) ratio."

Finally, there is the "loss of life expectancy." This risk is calculated by multiplying the individual lifetime risk by the average remaining lifetime (assuming 72 years as an average lifespan). Each of these measures provides a different perspective on the potential health impact of exposure to a carcinogen.

16.2 Risk Assessment of Endocrine Disrupters

The 1996 amendment to the Safe Drinking Water Act (SDWA) and the newly enacted Food Quality Protection Act (FQPA) mandated the EPA to establish a screening and testing program for endocrine disruptors, to be operational by September 2000. An advisory panel recommended testing 87,000 compounds for their potential to disrupt the endocrine system. Prior to testing, chemicals undergo an initial sorting process. The high-priority group includes all pesticides and compounds with an annual production volume exceeding 10,000 pounds. Due to the large number of chemicals, the screening procedure is divided into two tiers.

The first tier, or pre-screening, involves short-term tests designed to determine whether the chemical interacts with estrogen, androgen, or thyroid receptors. It consists of three in vitro assays and five in vivo assays in rodents, frogs, and fish. Compounds testing positive in the pre-screening move to the second tier, which includes five assays: a two-generation reproductive toxicity study in mammals, a reproductive toxicity study in birds, a fish life-cycle assay, a crustacean life-cycle assay, and an amphibian reproductive toxicity assay. Compounds showing endocrine-disrupting activity in the second-tier study undergo additional hazard assessment.

The estimated cost of the first-tier screening is $15 million and the second-tier screening is estimated at $25 million. This comprehensive screening and testing program aims to assess the potential endocrine-disrupting effects of a wide range of chemicals.

16.3 Ecological Risk Assessment

While human-health risk assessment has been in place for several years, a new concern has emerged regarding ecological risk assessment. Although government agencies require ecological risk assessment before implementing new projects, there is a growing recognition of the complexity of the issues involved.

Firstly, in contrast to human-health risk assessment, which focuses on individuals of a single species, ecological risk assessment deals with populations of thousands of species. Secondly, due to the complexity and interconnected nature of ecosystems, there is a challenge in selecting appropriate endpoints. For instance, the symbiotic relationship between freshwater mussels and fish illustrates this complexity, where the demise of a fish species can lead to the extinction of the mussel population.

Furthermore, there is a growing understanding that ecosystem degradation can result not only from chemical factors but also from biological and physical factors, such as the introduction of exotic species or land development. Even when considering chemicals alone, the multitude of chemical agents in the environment and their potential cumulative or synergistic effects make studying the adverse impact of a single chemical highly speculative.

In 1992, the EPA published the Framework for Ecological Risk Assessment. However, the EPA advisory body, the Risk Assessment Forum, is currently scrutinizing the field to establish the groundwork for new ecological risk assessment guidelines. The goal is to address the intricacies of assessing risks to ecosystems, considering the multitude of factors that can contribute to ecological changes.

16.4 Adaptive Management Strategies in Response to Changing Environmental Conditions

Adaptive management emerges as a crucial and dynamic approach in response to the ever-changing complexities of environmental conditions. At its core, this strategy embodies a flexible and iterative process, acknowledging the inherent uncertainties in environmental management. The fundamental principles of adaptive management emphasize the necessity of flexibility, allowing for the continuous adjustment of management plans as conditions evolve. It promotes a culture of continuous learning, urging stakeholders to glean insights from both successful and unsuccessful outcomes. Collaborative decision-making is integral, involving diverse stakeholders and fostering partnerships to incorporate a spectrum of perspectives and expertise. The iterative nature of adaptive management involves revisiting and updating strategies based on ongoing assessments, creating a dynamic and responsive framework. Within practical applications, examples abound across various domains. Ecosystem restoration projects, for instance, implement flexible, science-based plans that can be adjusted based on the success of native species and prevailing local conditions. In the face of climate change, adaptive strategies include modifying conservation plans in response to shifting migration patterns or altered habitat suitability. Water resource management embraces real-time data to allocate resources judiciously and adjusts release schedules based on ecological and human needs. Wildlife conservation efforts exemplify adaptive management by monitoring populations and modifying conservation measures in response to changing behaviors or circumstances. Despite its merits, challenges persist, such as robust data collection, effective stakeholder communication, and resource limitations. Moving forward, the integration of emerging technologies and advocating for adaptive management principles in environmental policies are essential considerations to navigate future uncertainties effectively. In essence, adaptive management stands as a pivotal strategy, aligning itself with the dynamic nature of environmental conditions and ensuring the sustainable stewardship of our ecosystems.

16.5 Use of Ecological Restoration to Mitigate the Impacts of Pollution

Ecological restoration serves as a powerful tool to mitigate the impacts of pollution, offering a holistic and sustainable approach to rejuvenating ecosystems affected by contaminants. Pollution, whether from industrial discharge, agricultural runoff, or other sources, can severely degrade environmental quality. Here is how ecological restoration plays a crucial role in mitigating these impacts:

1. **Soil Remediation:** Contaminated soils are a common consequence of pollution, often resulting from industrial activities or improper waste disposal. Ecological restoration techniques such as phytoremediation, which involves using plants to extract, accumulate, or degrade pollutants, can help remediate soil. Certain plant species can absorb and detoxify pollutants, thereby restoring soil health over time.

2. **Riparian Zone Restoration:** Pollution in water bodies, such as rivers and streams, can have detrimental effects on aquatic ecosystems. Restoring riparian zones, the areas along watercourses, helps mitigate water pollution. Planting native vegetation along riverbanks can stabilize soil, reduce erosion, and filter pollutants, improving water quality. This approach not only addresses the symptoms of pollution but also enhances the resilience of aquatic habitats.

3. **Wetland Restoration:** Wetlands act as natural filters, trapping and breaking down pollutants. However, they are often impacted by pollution. Ecological restoration efforts focus on re-establishing or enhancing wetlands to improve water quality. Restored wetlands can efficiently remove nutrients, sediments, and contaminants, acting as a buffer against pollutants before they enter downstream ecosystems.

4. **Biodiversity Enhancement:** Pollution can lead to the decline of biodiversity as certain species are more sensitive to contaminants.

Ecological restoration aims to reintroduce native plant and animal species that are adapted to the local conditions. By restoring biodiversity, ecosystems gain resilience and are better equipped to withstand and recover from pollution events.

5. Brownfield Restoration: Areas contaminated by industrial activities, known as brownfields, can undergo restoration to transform them into functional ecosystems. This process involves soil remediation, habitat restoration, and the reintroduction of native vegetation. Brownfield restoration not only mitigates pollution but also contributes to urban biodiversity and community well-being.

6. Forest Restoration: Air pollution, including pollutants like sulfur dioxide and nitrogen oxides, can have detrimental effects on forest ecosystems. Ecological restoration projects focus on replanting native tree species, improving soil health, and enhancing overall forest resilience. Healthy forests act as carbon sinks, contributing to air quality improvement.

7. Community Engagement and Education: Ecological restoration is not solely about physical interventions; it also involves engaging communities and educating them about sustainable practices. By raising awareness and involving local communities in restoration efforts, there is a greater likelihood of reducing pollution at its source through changed behaviors and responsible environmental practices.

In essence, ecological restoration serves as a multifaceted approach to mitigate the impacts of pollution by addressing its root causes and rehabilitating affected ecosystems. By integrating scientific knowledge, community participation, and sustainable practices, restoration initiatives contribute to the long-term health and resilience of the environment in the face of anthropogenic pressures.

16.6 Incorporation of Habitat Management in Polluted Ecosystems

The incorporation of habitat management in polluted ecosystems is a critical component of holistic environmental restoration efforts. Habitat management involves manipulating and enhancing the physical and biological features of an ecosystem to improve its overall health and resilience. In polluted ecosystems, where the impacts of contaminants may have far-reaching consequences, habitat management strategies are designed to address pollution while fostering sustainable, functioning habitats. Here are key approaches to incorporating habitat management in polluted ecosystems:

1. Riparian Buffer Zones: Establishing riparian buffer zones along water bodies is a fundamental habitat management strategy. These zones, characterized by native vegetation, act as a natural filter for pollutants entering waterways. The vegetation helps stabilize soil, reduce erosion, and absorb nutrients, sediment, and contaminants, preventing them from reaching the aquatic ecosystem.

2. Wetland Creation and Restoration: Wetlands are effective in trapping and filtering pollutants. Habitat management involves creating or restoring wetlands to enhance their capacity to absorb and break down contaminants. Constructing artificial wetlands or restoring degraded ones can significantly contribute to water quality improvement by promoting biological processes that mitigate pollution impacts.

3. Native Vegetation Restoration: The reintroduction or restoration of native vegetation is a fundamental aspect of habitat management in polluted ecosystems. Native plants are often better adapted to local conditions and may possess the ability to tolerate or even remediate certain pollutants. Restoring vegetation helps stabilize soil, prevent erosion, and enhance overall habitat structure.

4. Soil Amendments and Phytoremediation: Habitat management can involve the use of soil amendments and phytoremediation techniques to address soil pollution. Certain plants can absorb and detoxify pollutants, a process known as phytoremediation. The strategic planting of such species, along with soil amendments to enhance microbial activity, can help rehabilitate contaminated soils.

5. Habitat Connectivity and Corridor Creation: Establishing habitat connectivity through green corridors is essential for promoting biodiversity and enhancing ecosystem resilience in polluted landscapes. Habitat corridors facilitate the movement of species, allowing them to adapt to changing conditions. This connectivity is particularly crucial for the survival and migration of wildlife in fragmented or polluted environments.

6. Urban Greening and Brownfield Transformation: In urban areas affected by pollution, habitat management focuses on urban greening initiatives and transforming brownfields into green spaces. Creating green infrastructure, such as parks and green roofs, not only improves air quality but also provides habitats for diverse flora and fauna, contributing to overall ecological health.

7. Adaptive Management and Monitoring: Incorporating adaptive management principles is crucial in polluted ecosystems. Regular monitoring of habitat conditions and pollutant levels allows for adaptive responses and adjustments to management strategies. Adaptive management ensures that habitat interventions are effective and responsive to changing environmental conditions.

8. Community Engagement and Stewardship: Habitat management in polluted ecosystems involves engaging local communities in stewardship and restoration activities. Educating and involving communities in habitat restoration fosters a sense of ownership and responsibility, encouraging sustainable practices that contribute to pollution reduction and habitat preservation.

In conclusion, habitat management in polluted ecosystems is a dynamic and integrative approach that aims to restore ecological functions, enhance biodiversity, and mitigate the impacts of contaminants. By combining various strategies, from riparian buffer zones to adaptive management, habitat management contributes to the sustainable recovery of ecosystems affected by pollution.

16.7 Case Studies Highlighting Successful Restoration Projects

Several successful restoration projects around the world serve as compelling case studies, showcasing effective strategies for rehabilitating ecosystems and promoting biodiversity. Some examples are as follows:

1. Loess Plateau Watershed Rehabilitation Project (China): The Loess Plateau in China faced severe soil erosion and degradation due to unsustainable agricultural practices. The project, initiated in the late 1990s, employed a holistic approach combining afforestation, terracing, and sustainable land management practices. By involving local communities, the project successfully restored vegetation cover, reduced soil erosion, and improved water retention. The transformed landscape not only enhanced ecological resilience but also improved the livelihoods of communities dependent on the land.

2. The Everglades Restoration (United States): The Everglades, a unique wetland ecosystem in Florida, suffered from decades of water mismanagement and habitat loss. The Comprehensive Everglades Restoration Plan (CERP), launched in 2000, is an ambitious initiative to restore the natural flow of water, improve water quality, and rehabilitate habitats. With ongoing efforts to remove levees, construct reservoirs, and control invasive species, the project has shown positive outcomes in terms of re-establishing natural water patterns and supporting the recovery of native flora and fauna.

3. The Aral Sea Restoration (Central Asia): Once the fourth-largest lake globally, the Aral Sea experienced a drastic reduction in water volume due to diversion for irrigation projects. The World Bank-supported North Aral Sea project focused on building a dam to separate the North and South Aral Seas and restore water levels in the North Aral. As a result, the ecosystem began to recover, and fish stocks rebounded. The project exemplifies the potential for large-scale interventions to reverse environmental degradation.

4. Rewilding in the Netherlands: The Oostvaardersplassen nature reserve in the Netherlands serves as a rewilding success story. Initially, the area was artificially created but left to develop naturally with minimal human intervention. The project aimed to establish a self-regulating ecosystem with large herbivores like red deer and Heck cattle. Over time, the reserve evolved into a diverse wetland habitat, attracting a variety of bird species and other wildlife. The approach underscores the importance of allowing nature to take its course in certain restoration efforts.

5. Chesapeake Bay Restoration (United States): The Chesapeake Bay, the largest estuary in the United States, faced pollution and habitat degradation due to urbanization and agricultural runoff. The

Chesapeake Bay Program, a collaborative initiative involving federal and state agencies, implemented measures to reduce nutrient and sediment loads. The restoration plan included implementing agricultural best practices, upgrading wastewater treatment plants, and restoring riparian buffers. Over time, water quality has improved, and efforts continue to enhance the health of the bay and its surrounding ecosystems.

These case studies demonstrate that successful restoration projects require a combination of scientific expertise, community engagement, and sustained commitment. By learning from these examples, stakeholders can gain insights into effective approaches for addressing environmental challenges and restoring ecosystems to a more resilient and sustainable state.

16.8 Mitigating Ecotoxicological Risks

16.8.1 Environmental Regulations and Policies

16.8.1.1 Overview of international, national, and regional regulations

Environmental regulations and policies form the backbone of efforts to control and manage the impact of pollutants on ecosystems. These regulations operate at various levels, including international, national, and regional, and are designed to set standards, monitor compliance, and enforce measures to protect the environment.

1. International Regulations:

- Treaties and agreements addressing global environmental issues.
- Examples include the Stockholm Convention on Persistent Organic Pollutants (POPs) and the Basel Convention on the Control of Transboundary Movements of Hazardous Wastes and Their Disposal.

2. National Regulations:

- Laws enacted by individual countries to safeguard their environments.

- Examples include the Clean Water Act and the Clean Air Act in the United States, the Water Framework Directive in the European Union, and the Environmental Quality Act in Malaysia.

3. Regional Regulations:

- Collaborative efforts among neighboring countries to address shared environmental challenges.
- Examples include regional agreements on air quality control in Europe and water management in river basins.

16.8.2 Role of Regulatory Frameworks in Controlling Pollutant Discharges

1. Setting Standards and Limits:

- Establishing permissible levels of pollutants in air, water, and soil.
- Defining emission limits for industries and discharge standards for wastewater treatment.

2. Monitoring and Reporting:

- Implementing systems for monitoring pollutant levels in different environmental compartments.
- Requiring industries to report their emissions and discharges regularly.

3. Enforcement and Penalties:

- Establishing regulatory authorities responsible for enforcement.
- Imposing penalties on non-compliant entities to incentivize adherence to regulations.

16.8.3 Success Stories and Challenges in Implementing Ecotoxicology-Based Regulations

1. Success Stories:

- Reduction of ozone-depleting substances through the Montreal Protocol.
- Improvement in air and water quality in response to regulatory measures.

- Preservation of endangered species and habitats through environmental regulations.

2. Challenges:

- Enforcement gaps and regulatory loopholes lead to non-compliance.
- Balancing economic development with environmental protection.
- Addressing emerging pollutants not covered by existing regulations.

Environmental regulations and policies play a pivotal role in mitigating the impact of pollutants on ecosystems. While success stories demonstrate the effectiveness of well-implemented regulations, challenges persist, requiring continuous refinement and adaptation of regulatory frameworks. As the field of ecotoxicology evolves, policymakers must stay abreast of emerging issues to ensure that regulations remain effective in safeguarding the environment for current and future generations.

Bibliography

- Albert, A. (2012). *Selective Toxicity: The Physico-Chemical Basis of Therapy*. Springer Science & Business Media.
- Burcham, P. C. (2013). *An Introduction to Toxicology*. Springer Science & Business Media.
- DuBois, K. P., & Geiling, E. M. K. (1959). *Textbook of Toxicology.*
- Hodgson, E. (Ed.). (2011). *A Textbook of Modern Toxicology*. John Wiley & Sons.
- Hoffman, D. J., Rattner, B. A., Burton Jr, G. A., & Cairns Jr, J. (Eds.). (2002). *Handbook of Ecotoxicology*. CRC Press.
- Imlay, J. A., Chin, S. M., & Linn, S. (1988). Toxic DNA damage by hydrogen peroxide through the Fenton reaction in vivo and in vitro. *Science, 240*(4852), 640–642.
- Johansson, I., & Ingelman-Sundberg, M. (2011). Genetic polymorphism and toxicology—with emphasis on cytochrome p450. *Toxicological Sciences, 120*(1), 1–13.
- Klaassen, C. D., & Amdur, M. O. (Eds.). (2013). *Casarett and Doull's Toxicology: The Basic Science of Poisons* (Vol. 1236, pp. 189–190). New York: McGraw-Hill.
- Lee, B. M., Kacew, S., & Kim, H. S. (2017). *Lu's Basic Toxicology: Fundamentals, Target Organs, and Risk Assessment.* CRC Press.
- McCarthy, J. F., & Shugart, L. R. (Eds.). (1990). *Biomarkers of Environmental Contamination* (p. 457). Boca Raton, FL: Lewis Publishers.
- Morrow, P. E., Gibb, F. R., Gazioglu, K. M., Mahoney, E., Davies, H., Feingold, B., & Wood, D. (1967). A study of particulate clearance from the human lungs. *American Review of Respiratory Disease, 96*(6), 1209–1221.
- Odum, H. T. (1983). *Systems Ecology: An Introduction.*
- Pani, B. (2010). *Textbook of Toxicology*. IK International Pvt Ltd.

- Ratcliffe, D. A. (1967). Decrease in eggshell weight in certain birds of prey. *Nature, 215*(5097), 208–210.
- Roberts, S. M., James, R. C., & Williams, P. L. (Eds.). (2022). *Principles of Toxicology: Environmental and Industrial Applications.* John Wiley & Sons.
- Shelford, V. E. (1912). Ecological succession: V. Aspects of physiological classification. *The Biological Bulletin, 23*(6), 331–370.
- Tansley, A. G. (1935). The use and abuse of vegetational concepts and terms. *Ecology, 16*(3), 284–307.
- Walker, C. H., Sibly, R. M., & Peakall, D. B. (2005). *Principles of Ecotoxicology*. CRC Press.
- Willhite, C. C., & Smith, R. P. (1981). The role of cyanide liberation in the acute toxicity of aliphatic nitriles. *Toxicology and Applied Pharmacology, 59*(3), 589–602.
- Zakrzewski, S. F. (2002). *Environmental Toxicology*. Oxford University Press, USA.
- Zbinden, G., & Zbinden, G. (1976). The role of pathology in toxicity testing. In *Progress in Toxicology: Special Topics Volume 2* (pp. 8–18). Springer Berlin Heidelberg.

Index

www.ingramcontent.com/pod-product-compliance
Ingram Content Group UK Ltd.
Pitfield, Milton Keynes, MK11 3LW, UK
UKHW022159190926
475702UK00003B/105
* 9 7 8 9 8 1 5 1 2 9 6 8 7 *